WORMWOOD
FOREST

WORMWOOD
FOREST

a natural history of chernobyl

Mary Mycio

Joseph Henry Press
Washington, D.C.

Joseph Henry Press • 500 Fifth Street, NW • Washington, DC 20001

The Joseph Henry Press, an imprint of the National Academy Press, was created with the goal of making books on science, technology, and health more widely available to professionals and the public. Joseph Henry was one of the founders of the National Academy of Sciences and a leader in early American science.

Any opinions, findings, conclusions, or recommendations expressed in this volume are those of the author and do not necessarily reflect the views of the National Academy of Sciences or its affiliated institutions.

Library of Congress Cataloging-in-Publication Data

Mycio, Mary.
    Wormwood forest : a natural history of Chernobyl / Mary Mycio.
       p. cm.
    Includes index.
    ISBN 0-309-09430-5 (cloth)
    1. Radioisotopes—Environmental aspects—Ukraine—
Chornobyl Region. 2. Chernobyl Nuclear Accident, Chornobyl,
Ukraine, 1986—Environmental aspects. 3. Radioisotopes—Health
aspects—Ukraine—Chornobyl Region. I. Title.
QH543.5.M93 2005
577.27′7′094777—dc22

                                       2005012715

Cover design by Michele de la Menardiere; Radiation Damaged Tree © NOVOSTI/Science Photo Library.

Printed in the United States of America.

First Printing, August 2005
Second Printing, September 2005

To my parents

Contents

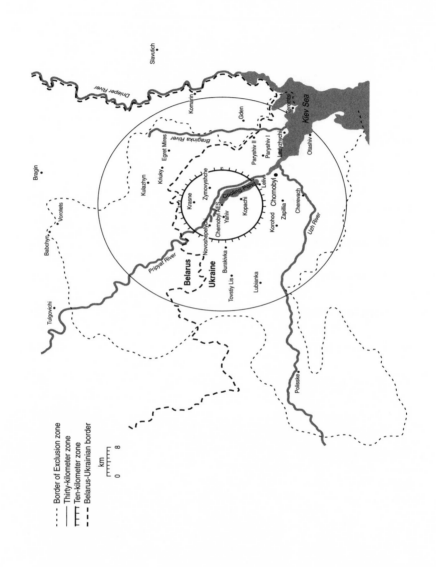

Border of Exclusion zone
Thirty-kilometer zone
Ten-kilometer zone
Belarus–Ukrainian border

km
0   8

Slavutich

Dnieper River

Bragin

Vorolets

Babchyn

Tulgovichi

Belarus
Ukraine

Pripyat River

Tovstiy Lis

Burakivka

Lubianka

Krasne

Novoshepelychi

Zymovyshche

Yaniv

Chernobyl AES

Cooling Pond

Kopachi

Korohod

Zapillia

Chornobyl

Cherevach

Uzh River

Polisske

Kulazhyn

Kriuky

Egret Mires

Braginka River

Komarin

Gden

Panyshiv II

Panyshiv I

Ladyzhichi

Teremtsi

Kiev Sea

Otashiv

Lelvik

Preface

When evidence of a mysterious Soviet nuclear incident leaked out of the radiation detectors of a Swedish nuclear station on April 28, 1986, I had just moved from New York to start my new job as an associate at a big Los Angeles law firm. Temporarily ensconced behind an absent partner's expansive desk while my own office was being made ready for me, I was working on some case files when the phone call came from New York. "A nuclear bomb exploded in Ukraine," exclaimed a friend who had not yet really forgiven me for leaving Manhattan. I don't recall how I expressed my shock, but after gleaning from her what little information she had, I dashed to the law firm's kitchen and tuned into CNN on the kitchen's TV set.

Western scientists quickly determined that the type of radiation that had invaded several European borders came not from a bomb but from a civilian nuclear power plant. Soon afterwards, the Soviet Union admitted that an accident on April 26 had destroyed part of what was, until then, the obscure Chernobyl atomic energy plant in the Soviet Ukrainian republic.

Watching television during working hours was not an activity that high-priced law firms appreciate in junior associates—especially ones that had just started their jobs. But the nuclear disaster riveted nearly everyone's attention, and my ethnic Ukrainian background made me a minor office celebrity for the 15 minutes it took for everyone else's minds to return to clients, cases, and billing. *My* mind never did—a fact that eventually became evident to the law firm's partners. As one of them put it before nicely suggesting I consider a different career, I lacked a legal "fire in my belly." He was absolutely right. I had a fire in my belly, but it was not a legal fire. It was an atomic fire—a determination to find out the truth about Chernobyl.

Uncovering that truth proved to be a daunting task, at least in the beginning. Unlike the disaster of the space shuttle *Challenger*, which had exploded on television screens just a few months earlier, the public saw no live, dramatic images of Chernobyl, just some grainy photos of the blackened crater in the ruined fourth reactor block. But that was merely a building, like a charred piece of furniture, and not an object to evoke horror, sympathy, or even much anxiety. It was static and far away, unlike the radioactive cloud that drifted around the northern hemisphere, prompting L.A. disc jockeys to joke about buying Chernobyl umbrellas when the cloud hit Southern California in early May.

There weren't even many voices to humanize the dry pronouncements of official Soviet news agencies and scientists. In the days before the Internet, only an occasional letter or ham radio operator could break through the impenetrable Iron Curtain to provide some uncensored news about the disaster. Well, "news" might be too kind a word. It was mostly rumor, speculation, and science fiction, like a wacky story in some supermarket tabloid about a six-foot-tall Chernobyl chicken. I didn't believe it, of course, but I duly clipped it and placed it in my fattening research files.

Like the radiation itself, the Chernobyl story was also invisible, rendered so by the habitual Soviet secrecy that even Mikhail Gorbachev's glasnost policy did not initially overcome. But even when glasnost did emerge victorious, the traditional secrecy of the international nuclear industry took over. Thus, when Soviet scientists met their Western counterparts at an unprecedented meeting to discuss the disaster in August 1986 in Vienna, it seemed that both sides operated according to an unspoken pact to minimize the disaster's effects. They told the truth, but they buried it so deep in the footnotes of scientific reports and the jargon of obscure journals that almost none of it could emerge to penetrate public consciousness.

After quitting full-time law in 1988, I spent a year as a bicoastal Chernobyl junkie, spending nearly all of my time in libraries in New York and Los Angeles and supporting my habit with an occasional short-term legal job. My favorite place was the restricted library at Brookhaven National Laboratory, where I snuck in by means I can no longer remember. The copying machines were free, and I could sit undisturbed for hours finding tidbits about real Chernobyl oddities like the "Minsk shoe" I mention in Chapter 1. I had hoped to write a book and came very close to a publishing contract but it fell through. With the clarity of hindsight, however, I am glad it did. Otherwise my second book on the subject would largely have to have been a retraction. Because all the while, I was looking for lies to expose. No bit of information existed merely as a fact but as a clue to a deep underlying truth that would reveal a massive cover-up by both East and West. My Ukrainian-American upbringing had instilled a visceral distrust of the Soviet Union in me. Hollywood scenarios such as the *China Syndrome* and *Silkwood* also made me wary of the Western nuclear industry. Some of that skepticism was justified. The Soviets' delay in announcing the simple precautions to take against radioactive iodine were unforgivable. So was the West's initial refusal to recognize the resulting thyroid cancer epidemic that became evident five years later. On many other issues, however, both sides told the truth—albeit in doses and over many years. Some truths did not become known until the USSR collapsed in 1991.

Reasonable minds may differ about the value of those truths. Initially, the disaster made me (and, I'm sure, many other people) oppose nuclear energy. In 1986 that was a painless position to hold, because the price of American dependence on foreign oil had not yet become two Iraq wars, the second of which still has undetermined costs and consequences. Nor had I yet moved to Ukraine, whose complete dependence on Russian fossil fuels seriously compromised the young state's political independence. It was also before I could feel the real evidence of global warming on my own skin.

For the record, I have gone from adamant opponent of nuclear energy to ambivalent supporter—at least for giving a window of time for reducing our dependence on fossil fuels while pursuing research on alternative energy sources. But even those alternatives can have environmental costs. For example, to harness the energy of Ukraine's Dnieper River, Europe's third largest, over the years the Soviets transformed it into a series of shallow, eutrophic reservoirs where fish perish by the thousands during hot summers. Though there is probably less of a downside to wind or solar energy, it seems there is little we can do to feed the world's growing appetite for energy without doing some damage. It is a choice of lesser evils.

Before Chernobyl, the odds of a reactor meltdown at one of the world's 300 or so nuclear reactors was considered so miniscule it was practically disregarded. Chernobyl changed the odds to one meltdown every 30 years. Is that too high a price to pay? The extraordinary and unexpected fate of the evacuated "zone of alienation" around Chernobyl provides only a part of the answer. I hope that the rest will form in the mind of the reader after joining me on my journeys through the fascinating, beautiful—and radioactive—Wormwood Forest.

I write this in the wake of Ukraine's Orange Revolution—a witty, peaceful, and joyous uprising that swept authoritarian President Leonid D. Kuchma out of power and ushered opposition leader Viktor A. Yushchenko in to replace him. The dioxin poisoning that scarred and cratered the new president's face is mute testimony to the old regime's diabolical methods. But it is my fervent hope that Yushchenko's disfigurement will be like the radioactive poisoning of Chernobyl: a terrible wound that fades with time, leaving a warning about the past—and an abiding hope for the future.

Kiev, Ukraine
September 2005

1

Wormwood

In the years since the 1986 Chernobyl nuclear disaster spewed radiation around the globe and smudged the map of the then-Soviet Union with heavy contamination, the very word "Chernobyl" has become a synonym for "horrific disaster," conjuring the frightful radioactive deserts that landscape Atomic Age science fiction and resonate deeply in modern imaginations haunted by the specter of nuclear war.

Surely, whenever I thought about the irradiated lands 50 miles north of Kiev, it was like contemplating a black hole. All I could picture was a dead zone, like a giant parking lot paved with asphalt or a barren desert of dust and ash where nothing could grow and nothing living could survive without protective gear. Only gloomy shades of black and gray colored my mental images.

But when I first visited the Chernobyl region, 10 years after the disaster, I was surprised to find that the dominant color was green. My notes from that trip are filled with emphatically underlined and circled comments like "feral fields," "forests," and "wildlife?!" Contrary to the myths and imagery, Chernobyl's land had become a unique, new eco-system. Defying the gloomiest predictions, it had come back to life as Europe's largest nature sanctuary, teeming with wildlife. Like the forests, fields, and swamps of their unexpectedly inviting habitat, the animals are all radioactive. To the astonishment of just about everyone, they are also thriving.

But to appreciate the land's extraordinary resurrection, you first have to understand its demise.

Pripyat's old wedding registry was not easy to find, even though we had the address. Despite the barbed wire perimeter around this radio-active ghost town less than two miles from the Chernobyl power plant, the hull of an empty high-rise on Friendship of Nations Street had been emptied of anything with the slightest value, including most of the metal signs that announced shops and services. Former residents and looters had stripped apartments and offices down to their faded wallpaper. Only one empty room hinted at its former function, probably because the cardboard sign on the door, reading "School of Communist Labor," was too worthless to steal. But there was no sign of a wedding registry.

A perplexed Rimma Kyselytsia led our little group of explorers outside into a small square surrounded by empty apartment buildings. She studied the number painted on the side of the building and shook her blond curls in confusion. "It's the right address. So, where is it?"

Since Rimma was the guide, my other companion—a botanist named Svitlana Bidna—and I shrugged helplessly.

My dosimeter beeped slowly. The radiation monitor's liquid crystal screen displayed 80 microroentgens an hour. That was several times normal background levels, which range from 15 to 25 in most places. Decontamination, rain, and time have long since washed off much of the radioactive grime that coated the town after Chernobyl's fourth reactor exploded in the wee hours of April 26, 1986. Pripyat was the plant's bedroom community. Heralded as the world's youngest city when it opened its doors in the mid-1970s, Pripyat also turned out to be its shortest lived.

A short flight of concrete stairs sprouting saplings and moss led to the back of the building where Rimma explored a row of what seemed to have once been stores and offices. The glass storefronts were all shattered, exposing the bare rooms to the elements, and she quickly spied a faded red carpet runner, lying dirty and twisted with shards of glass, plaster, and deep piles of yellowed paper.

"This is it!" she exclaimed, vindicated in her guiding skill. Rimma was a Tatar with aquarium eyes and a matter-of-fact but realistic attitude towards her radioactive workplace.

The red carpet runner once led couples to secular Soviet marriage in the Pripyat Registry of Citizens' Civil Status. Known by its Ukrainian acronym as a *ZAHS (ZAGS* in Russian), the office was not merely a marriage registry. *ZAHSs* documented the legal passages in Soviet citizens' personal lives from cradle to grave, issuing birth certificates and death certificates and everything in between.

The deep piles of brittle paper on the floor were *ZAHS* forms and applications. Ivory cards informing brides and grooms of their wedding dates were mixed up with spilled stacks of divorce applications and forms to apply for "compensation in the form of gold wedding rings." Hanging lopsided on the back wall, a red-lettered cardboard sign exhorted newlyweds: "Stand on the threshold of your introduction to the deep familial and social traditions of the Soviet people."

In the neighboring room, two tall bookshelves had toppled over, spilling dozens of pulp folders containing the *ZAHS* archives into a moldy pile. Although the 1986 archives were missing, the records went back to the early 1970s, when the town first opened its doors. Judging by one fat folder, many couples applied to cut to the front of the wedding queue because they had already had a child together.

Sixteen marriage ceremonies took place on the last full day of human life in Pripyat. The only public record of those nuptials, tinged in hindsight with so much sadness, can be viewed in a five-minute film at Kiev's Chernobyl museum. The split-second scene of the bride and groom leaving the storefront wedding registry is too fleeting to see their expressions, but the point of the wedding in the silent and grainy film was to show that April 26, 1986, was an ordinary, if unusually warm, Saturday afternoon in Pripyat. Oblivious to the radioactive cloud invisibly blanketing them, couples wheeled infants in strollers. Toddlers in shorts kicked a ball around a dirt playground. Women in sleeveless summer dresses gathered outdoors under a vendor's umbrella, in the large groups that always signified something (anything) being sold in the shortage-ridden Soviet Union.

But the anonymous KGB cameraman knew that something was wrong. Gamma rays left flashes of light on the scene he filmed of two men in camouflage and gas masks nodding to an unprotected and obviously surprised civilian. Armored personnel carriers drove down Pripyat's boulevards, while uniformed officers checked radiation on a truck's tires. Water trucks washed the streets with foamy detergent, leaving puddles in which sparrows splashed. From the roof of a Pripyat high-rise, the cameraman filmed the Chernobyl plant, shrouded in such a thick cloud of smoke and haze that only its dim outline was visible.

What did not get recorded on film was the nighttime explosion that ripped through the Number 4 reactor complex, spewing flames, sparks, and chunks of burning radioactive material into the air and, subsequently, around the northern hemisphere. Red-hot pieces of nuclear fuel and graphite fell on the roof, starting 30 fires and causing the roof to collapse into the reactor hall. By dawn the roof fires had been put out by 37 fire crews working without protection or dosimeters. Many became ill with acute radiation sickness. Thirty-one died, but at that point no one knew that the explosion had completely exposed the reactor core. The government commission from Moscow didn't arrive until Saturday night, and it wasn't until Sunday morning that its members could helicopter over the cavernous hulk to see that the explosion had ignited an extremely intense graphite fire. The graphite fire was releasing millions of curies of radioactivity that lit the air above the ruined reactor with an eerie glow. The crisis was actually an unprecedented disaster and it was far from over.

That morning, in his apartment not far from the *ZAHS* office, Volodymyr Pasichnyk had been watching his teenaged son playing with the dial on the TV set when the receiver suddenly tuned in on an odd frequency. "There was no picture, just talk, probably by walkie-talkie," he recalled when I talked with him 15 years later. "They were talking about 'people in hospitals' and 'hundreds of buses to Pripyat.'" Like most people in Pripyat—all of whose lives were somehow connected with the plant—he had heard rumors about something bad at the fourth reactor block. At that moment he understood the enormity of what had happened. The town was being evacuated.

The official announcement came on Pripyat radios at 10 o'clock Sunday morning. In only four hours, beginning at 2:00 p.m., 1,100 empty buses drove into Pripyat and drove out with nearly all of the town's 45,000 residents in a convoy that was more than 10 miles long. It was, the authorities said, only meant to be for three days. Perhaps they really meant it. But that was before anyone knew that the graphite fire would melt the fuel and belch the daily equivalent of several Hiroshima bombs for 10 full days, altogether releasing five times as much radioactivity as the initial explosion.

Pripyat could never be inhabited again. The ritual human cycles of birth and marriage, divorce and death, recorded by *ZAHS* scribes in the thick pulpy ledgers of Soviet bureaucracy, ended on April 26, 1986. And the stacks of cards for secular baptisms, with spaces for the names of new Soviet citizens, will never be filled in.

BIBLICAL BOTANY

Svitlana Bidna, my botanist companion, walked with Rimma and me on the thick moss carpets strewn over Pripyat's crumbling asphalt roads, giving the tangled overgrowth of vegetation names and clarity. The straight rows of poplars lining the streets were planted when the town was built. The asters, still blooming in late October, were once garden flowers that enlivened the cinder-block sterility of the new Soviet town. But poplars are now growing out of storefronts and stairwells. Asters have taken to the wild in large lavender fields. And a diverse profusion of wild species are filling the cracks in the concrete and the voids left by people. Svitlana, who has been studying the town's transmutation to forest after Chernobyl, predicts that the buildings will stand half a century, perhaps. Though, if left to itself, the greenery will consume most of the asphalt roads and concrete plazas in another decade or so.

Lichens such as the bright orange *Xanthoria* secrete chemicals that destroy the crystalline structure of minerals in concrete. Acids in the mosses that grow on dead leaves soften asphalt, crumbling it into pebbles. Birch, maple, and pine trees sprout from the cracks, buckling pavements with their roots and exposing more crevices for greenery to grow. Reeds grow in patches where the shallow water table is recreating once-drained swamps. A forest of silver birch, willow, wild pear, and pine fills the former soccer field, and islands of grass covering broad concrete plazas sprout tall bushes of false indigo, its long flower clusters dried into black tassels.

Pripyat was coming to resemble one of those fabled lost cities, devoured by jungle. Abandonment echoed in every corner of the crumbling monument to the disaster. A Ferris wheel and bumper cars rust away in a tiny amusement park, scheduled to open on May 1, 1986, and never used. The town pool's three-story glass facade had completely shattered into deep piles mixed with tiles from the wall mosaics that resembled incomplete jigsaw puzzles, with chalky green lichen replacing the missing pieces and making it impossible to tell what they once depicted. Shrubs and saplings grew in kindergartens scattered with tiny shoes, broken toys, and heartbreakingly small gas masks. We climbed up a high-rise where Svitlana showed us a good-sized birch tree growing from the center of a kitchen emptied of everything but an overturned table.

"Pripyat began returning to nature as soon as the people left, and there was no one to trim and prune and weed," said Svitlana as we started heading back to our car. "It takes a lot of human effort to maintain urban landscapes."

Back near the wedding registry, Rimma crouched down to a short bush that had grown out of a crack between the road and the curb. It was about a foot tall, with small cottony flowers growing directly from purplish stems.

She pulled off one of the leaves and crushed it between her fingers for me to sniff the unpleasant, varnishy aroma, reminiscent of shoe polish.

"What is it?" I asked, wrinkling my nose.

"*Chernobyl*," she said, using the common—but incorrect—pronunciation. In fact, chernobyl with an "e" is the Russianized version of the Ukrainian word *chornobyl*. You won't find *chernobyl* or *chornobyl* in most Russian dictionaries, except in reference to the disaster, although the word *chernobyl'nyk* is used in some Russian regions in reference to the herb. But because the first version has become the commonly accepted spelling for the disaster and the nuclear station, I will use *Chernobyl*, with an "e," to refer to them. I will use *Chornobyl*, with an "o," to refer to the herb and the town.

"That's wormwood, right?" I asked, hoping to finally clarify the botanical question at the heart of the Chernobyl disaster's putative biblical symbolism. It is often said that the meaning of the Ukrainian word *chornobyl* is "wormwood," and the suggestion that the disaster fulfilled the biblical prophecy of the Wormwood star that augured Armageddon resonated deeply with the fear of nuclear apocalypse. But the botany was actually more complex.

Svitlana took a closer look at the plant and shook her head. "No, '*chornobyl*' is *Artemisia vulgaris.* 'Wormwood' is *Artemisia absinthium.* The Ukrainian common name is *polyn,*" she said, handing me a leaf from a different plant that looked much like *A. vulgaris,* except it was covered with fine silky hairs that gave it a whitish tinge. As I looked around, I noticed that the plants were everywhere.

I crushed it to release the volatile oil, much more pungent than the first plant.

Botanically and chemically, *Absinthium vulgaris* is so similar to *A. absinthium* that *A. vulgaris* is also sometimes called "wormwood," though "mugwort" is a more common English name. In Ukrainian, as well, *polyn* and *chornobyl* are sometimes used synonymously. Both plants are hardy perennials, tolerant of poor soil and thus plentiful in the sandy lands of the Polissia region—where the twelfth-century town of Chornobyl took its name from the plant and, in turn, gave it to the twentieth-century nuclear station seven miles away. Both are bitter medicinal herbs and natural pest repellants, ridding fleas from the home, slugs from the garden, and worms from the body. And both get their pungent fragrance from thujone, an organic toxin thought to be the psychoactive agent in absinthe, the infamous wormwood liqueur banned by most Western countries a century ago. Absinthe was said to produce an unusual intoxication and was highly addictive, although modern skeptics contend that the "high" and the habit most probably came from drinking the 75 percent alcohol absinthe required to dissolve the thujone and prevent it from clouding the emerald solution.

But if the thujone in *Artemisia vulgaris* is dilute, it is concentrated in *A. absinthium.* A crushed leaf of *polyn*-wormwood is much more pungent than a crushed leaf of *chornobyl*-mugwort. It is also more bitter and much more toxic, which is why animals happily nibble mugwort but leave wormwood alone. Even other plants avoid it. *A. absinthium*'s extremely bitter chemicals wash off the leaves and into the soil, poisoning it for other plants.

Given its natural repellant properties, many folk believed wormwood to have supernatural banishing powers. Mugwort, too, has magical properties, though none so potent. In Ukrainian folklore, both plants ward off the seductive and dangerous water nymphs called *rusalkas,* who lured victims with beautiful songs and then tickled them to death in crystal underwater lairs.

In Christian legend, when the biblical serpent was expelled from Eden, wormwood sprang in its trail to prevent its return. Indeed, the herb is a frequent biblical symbol for bitterness, calamity, and sorrow; its use to name the third sign of the apocalypse that opened this chapter conjured the desolation that would follow the apocalypse.

In the wake of the Chernobyl explosion, few people in the officially atheist Soviet Union had Ukrainian-language Bibles. But some of those who did noted that the word "wormwood" in the Wormwood star of the book of Revelation was translated as *polyn*—and was a very close botanical cousin to *chornobyl*. Suddenly, the biblical prophecy seemed to acquire new meaning: wormwood was radiation, and it presaged the nuclear apocalypse that would end the world. The story spread like wildfire through the notorious Soviet rumor mill and as far as Washington, D.C., where President Ronald Reagan was said to have believed it, too.

I first learned of the apocalyptic connection about a month after the disaster, when a Ukrainian friend in Poland wrote me about it in a letter. I had just moved to Los Angeles from New York City in what was supposed to have been my "American experiment." I didn't want to sever my Ukrainian-American roots entirely. But I did want to try living without the sometimes suffocating support of the ethnic ghetto that was an integral part of my life in Manhattan. Chernobyl put an end to that experiment before it even started. I recall crying on the phone with my best friend in New York and realizing that only someone with Ukrainian roots could share the pain I felt contemplating the swirl of televised speculation about the disaster's calamitous effects on a land that I had been raised to believe was very important to me but one that I had never seen because it was shrouded by an impenetrable Iron Curtain.

Chernobyl's putative apocalyptic connection became so widespread, combining fears of radiation with apocalyptic dread, that the state-controlled Soviet media took the highly unusual step of running interviews with leaders of the Russian Orthodox Church (the most tolerated religion in the USSR) to debunk it, largely by arguing that no man could know when the end of time was near.

Perhaps their arguments would have been better served by botany. Aside from the fact that *polyn* and *chornobyl* are different species of *Artemisia*, it is unlikely that the wormwood in Revelation referred to either of them. *Artemisia judaica* is widely cited as the most likely candidate for biblical wormwood.

But judging by a cursory perusal of the 975 results that an Internet search of "chernobyl wormwood" turned up, Armageddon-watchers seem untroubled by such technicalities (though they are troubled by others, such as how anyone can prove that "a third of the waters were made bitter" as predicated by Revelation 8-10). For them, Chernobyl equals wormwood, and the end of the earth as we know it is near. Far be it for me to dismiss biblical prophesy, but as we left the crumbling and *Artemisia*-choked landscape of Pripyat, it seemed that the only end Chernobyl heralded for certain was that of the Soviet Union.

NUCLEAR POWER

Back in the hired Soviet sedan driving on the sole road from Pripyat, the dosimeter perched on my knee began beeping faster as we drove due west of the power plant. The liquid crystal numbers suddenly started growing rapidly: 135, 165, 170, 228, 485, maxing out at the oddly apocalyptic 666 microroentgens when the car stopped under a tall white pillar topped with a red triangle.

The Soviet-era sign read: "V. I. Lenin Chernobyl Atomic Energy Station," but I was stumped as to the monument's possible meaning.

"It's a torch," Rimma explained. "It symbolizes the light that the Chernobyl plant produced."

The Wormwood star also blazed like a torch, I recalled, although biblical symbolism doesn't get you far in understanding how the disaster happened. What you need is a short detour into nuclear fundamentals.

Nuclear reactors like Chernobyl make light much like all thermal power plants. A reaction releases energy in the form of heat, which boils water to create steam, which then turns turbogenerators to produce electricity. In conventional plants, the reaction is the chemical combustion, or burning, of complex hydrocarbon molecules in fossil fuels. Combustion reactions involve chemical bonds, which join atoms together in molecules through the sharing of electrons, the quantum clouds of negative charge in the atoms' outer reaches. Heat breaks the bonds holding carbon and hydrogen atoms together in the hydrocarbons.

But the thing about most atoms is that their outer electrons are promiscuous when unattached and eager to bond with any chemically suitable atom that dallies nearby. That's why the world is not an invisible mist of loose atoms like the noble gases. When the hydrocarbons' bonds are broken, the freed carbon and hydrogen can't just be left dangling on their own, so they react with whatever is nearby, namely, the oxygen in the air. This creates simpler molecules, such as carbon dioxide, that need much less bond energy than the original hydrocarbon to stick together and are the main sources of the greenhouse gases that contribute to global warming. The excess chemical energy is released as heat (and light).

Instead of releasing energy by changing the electron bonds *between* atoms to create new and different molecules, the nuclear reaction releases energy by rearranging the nuclei *inside* atoms of one element to create atoms of a different element (Figure 1).

The strong nuclear force—the most powerful force in the universe—tightly binds the neutrally charged neutrons and positively charged protons inside the nucleus, overwhelming the mutual electrical repulsion between the protons. Operating on only the sub-subatomic level of quarks, the whimsically named points of energy that exist only in triplets (never alone) to make protons and neutrons, the

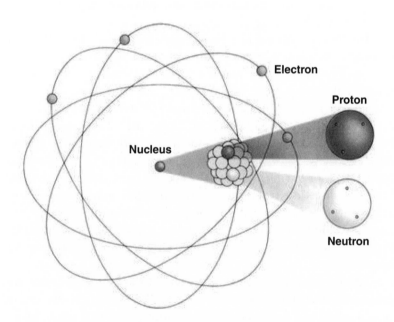

FIGURE 1 Diagram of an atom.

strong force glues the quarks together with an asymptotic strength that becomes stronger, like elastic, the farther apart they are pulled.

In nearly all of the matter we perceive with our senses, the strong force wins the tug of war with the repellant force between the protons, keeping the nuclei stable. And while it is a marvel of universal construction in and of itself, a stable atomic nucleus doesn't actually do much of anything except hold on to the electrons that do the chemical work of building matter.

The positively charged protons inside the nucleus do that job, attracting the negatively charged electrons. The protons also determine what element the nucleus will be. Elements are distinguished by the number of protons in the nucleus, which are reflected in the "atomic numbers" in the periodic table. As the number of protons increases, from hydrogen with one proton all the way to the man-made elements whose atomic numbers have passed 110, the repulsion between them increases. And the number of quarks, provided by neutrons, that the strong force needs to keep the nucleus glued together increases as well. For helium, with two protons, two neutrons suffice for balance. But the brittle white metal bismuth needs 126 neutrons to hold its 83 protons together. Bismuth, in fact, has the largest stable nucleus in the periodic table. All larger nuclei are unstable, although some smaller nuclei are too because their proton-neutron ratios are askew. Unstable nuclei have too many conflicting forces inside, and to achieve a semblance of internal peace, they must eventually decay, casting out their excess energy in the form of subatomic particles and gamma rays. It is this casting out of energy that is called "radiation."

Like all elements, uranium—the heaviest atom in nature—has several forms called isotopes, which share the same atomic number of 92 protons but vary in their numbers of neutrons and, hence, differ in their atomic weights and their stability. Natural uranium is a mixture of two isotopes. Uranium with an atomic weight of 238, designated U-238, makes up 99.3 percent. It is slightly radioactive and decays very slowly.

The remaining natural uranium, U-235, displays a unique property. Because of the particular instability of its 92 protons and 143 neutrons, uranium-235 can spontaneously fission—splitting apart into two or more nuclei of smaller elements plus two or three free neutrons. Fission produces dozens of different elements, but with the exception of plutonium, the daughter nuclei—called fission products—do not fission. They are, however, far more radioactive than the original uranium, which is why nuclear waste is much more dangerous than fresh fuel.

Fission is incomprehensibly fast. The nucleus splits less than 0.00000000000001 seconds after absorbing a neutron. Fission is also extremely energetic. In one of the more dramatic manifestations of Einstein's formula $E = mc^2$, it releases the inconceivably powerful strong force that had been holding the unstable nucleus together to begin with. It is so energetic that the nuclear energy released by fission in one gram of uranium fuel is the equivalent of burning three tons of coal.

The U-235 atoms in nature are too few for the occasional spontaneous fissions to do more than add marginally to uranium's heat—with one known geological exception. About 1.7 million years ago, in a place now called Oklo in the tiny West African country of Gabon, Earth itself ignited a natural fission chain reaction that consumed six tons of uranium over hundreds of thousands of years before burning itself out.

If uranium-235 occurs in higher concentrations, such as Oklo's three percent, fission neutrons are more likely to hit other U-235 nuclei, causing new fissions that produce more neutrons and more fissions in a self-sustaining chain reaction. But aside from more concentrated U-235, fission also needs something called a moderator, which slows down the neutrons. Fast neutrons make fission fizzle out.

It seems counterintuitive that slow neutrons are more effective at fission because we imagine them hitting the nucleus hard and fast, like a marble breaking an egg. A slowly rolled marble would tap the egg but not necessarily break it. Yet in the quirky quantum world of the neutron, the marble can magically slip inside the egg without breaking the shell. The shell breaks only *after* the marble is inside. Fast neutrons zip by too quickly to perform that trick.

Fission powers all nuclear reactors, although their designs differ in the materials used to slow the neutrons. Most of the world's nuclear reactors are water moderated. So was the Oklo natural reactor, which was saturated with groundwater.

Chernobyl and other reactors of its type in the former Soviet Union (called RBMKs, which stands for "reactor high-power boiling channel type" in Russian) were moderated with graphite—the crystalline carbon used in pencils. Schematically, the reactor core resembled a collection of giant pencils interspersed with toothpicks and thin straws. The pencils were the 2,488 graphite columns. The toothpicks were 1,660 thin zirconium and steel fuel channels containing about 200 tons of slightly enriched uranium. The straws were pipes inside the fuel rods for the cooling water that flowed through and boiled, taking away heat. Viewed from above, the core looked like a giant round Scrabble board of multicolored squares and tiles.

In water-moderated reactors, losing or decreasing the amount of cooling water stops fission because the neutrons speed up too much and the chain reaction stops. In graphite-moderated reactors, the graphite keeps moderating even if the flow of water slows or is lost. Fission continues and without sufficient water to cool it, the reactor runs faster and hotter.

This, by all accounts, is what happened in the early hours of April 26, 1986.

In a mangled and ill-advised experiment that violated every rule in the plant's own safety book, the pumps that powered Chernobyl's emergency water cooling systems were deliberately shut down. Without the power of the pumps, the water that normally flowed at a rate of 28 tons an hour slowed down, letting it absorb more heat from the fuel. Instead of simply boiling, it turned to steam. Within minutes, the enormous pressure from the steam exploded the core (Plate 1).

RADIOACTIVE RAIN

The initial explosion broke apart the graphite and fuel with the force of about 30 to 40 tons of TNT, casting the heaviest and most deadly debris directly outside the reactor. On radiation maps, which depict increasing levels of radioactivity with a palette progressing from green to yellow to orange and increasingly darker shades of red, the patches from the explosion are a deep brown and look like lopsided horns. One horn is a long, thin arrow about a mile wide that reaches 6 miles to the west before gradually fading to lighter colors as it extends for a full 60 miles. The other horn is a short, wider lobe that fell northwest, over the Pripyat River. Between them is the town of Pripyat, colored a lighter shade of red. Luckily, the lethal fallout fell around the town and not on it. Even more luckily, it wasn't raining in the reactor's immediate vicinity or the death toll might have been much higher. In the air the fallout dispersed and diluted, much like a cloud of smoke. Rain would have brought it down at deadlier concentrations, as it did in patches of Belarus that are as brown as the horns by the reactor. But those were—again luckily—in sparsely populated forests.

Located half a mile away from the concrete shelter built over the ruined fourth reactor, the Chernobyl torch monument stood directly in the narrow path of the long western arrow, known as the Western Trace in the scientific literature. The dosimeter's beeping increased as our little group got out of the metal-shielded car and followed Rimma onto a sandy field of moss, short grasses, and pine trees that stretched west of the tall stone pillar.

With one eye on my feet as we trudged through rugged terrain, I watched the liquid crystal numbers change: 800, 1,230, 547 microroentgens an hour. Then suddenly the screen went blank.

I showed the dosimeter to Rimma, who had one just like it. With a brief look at the empty screen, she explained: "It maxes out at two milli, and shuts down."

"Milli," or a thousandth, was the prefix on the special radiation unit—the roentgen—that measured the gamma rays in the air we were walking through. In translation, Rimma told me that the dosimeter shut down when exposure levels exceed two thousandths of a roentgen an hour.

Background radiation up to dozens of times the usual levels elsewhere is the norm in Chernobyl lands after 15 years, with an average of 43 microroentgens an hour in 2004. But these levels are usually counted in micros—or millionths of a roentgen. Walking around micro background, even many times normal, was comparable to the gamma radiation exposure of living in Denver. Milli-land was a different league, more like the black, naturally radioactive sands of the Brazilian coast.

"Should we really be here?" I asked uneasily, watching the dosimeter's numbers disappear and reappear depending on where we walked, once flashing 1,546 microroentgens—or one and a half milliroentgens—before blanking out again. Despite the uniform blocks of color on the radiation maps, the contamination was uneven and patchy.

My companions laughed. "A few minutes won't make a big difference," said Svitlana, the botanist, her face betraying nary a trace of concern.

Whether it would or wouldn't make a difference, I would never know. Radiation is of the random and bizarre world of quantum mechanics, and the health impact of low doses is hugely controversial in radiology circles. I shrugged and followed my companions, reminding myself that the black sands of Brazil are a tourist attraction and that three milliroentgens an hour really wasn't that much.

It is useful at this point to get some perspective on what these numbers mean. All of us are constantly exposed to natural background radiation from cosmic rays and radioactive elements, such as radon and potassium, in the earth. But some things we do can increase our dose from these natural sources. Air travel is one of them because the Earth's atmosphere is what protects us from cosmic rays and planes fly where the atmosphere is thinner and less protective. In fact, of all professions, astronauts and airline crews have among the highest levels of occupational radiation exposure, even higher than nuclear industry workers. The higher the altitude, the thinner the atmosphere is and the greater the exposure to cosmic rays, so the exact figures vary. But a New York to Paris round-trip will give you a radiation exposure of about three milliroentgens. A round-trip from New York to Los Angeles will give you a little less: a total exposure of two milliroentgens.

On April 30, 1986, radiation levels in the field we walked through were as high as 30 to 40 roentgens an hour, capable of inducing acute radiation syndrome within a few hours of exposure and death, for most people, in a day. In the immediate vicinity of the reactor, radiation levels were at least 200 roentgens an hour. I say "at least" 200 because the only radiation meters available at the time maxed out at 200 roentgens, in the same way that my dosimeter maxed out at two milliroentgens. To this day, no one knows what the highest radiation levels were in the first postdisaster days.

What is known is that of the 600 or so emergency workers that battled the reactor, 500 were hospitalized. Of these, 237 were suspected of having acute radiation syndrome, with its symptoms of nausea, diarrhea, hair loss, and skin damage, but only 134 cases of radiation sickness were confirmed. Among them were some of the bus drivers who drove from Kiev to Pripyat on Saturday night and Sunday morning knowing nothing about the radioactive danger. While they waited for the evacuation order, they hung around on the streets, played soccer, stood on the roofs of their buses to get a better view of the fire burning at the power plant, and even baked potatoes.

According to secret documents declassified by the Ukrainian government, in Pripyat, which sat between the two lethal lobes, gamma radiation measured 300 milliroentgens an hour on April 30. In the town of Chornobyl, the hourly exposure was 3 to 10 milliroentgens. But because roentgens measure only gamma rays (and X rays), while alpha and beta radiation also invisibly snap-crackle-popped the air and ground, actual radiation levels were higher.

Gamma radiation is simply easier to measure. A gamma ray is a photon—a packet of electromagnetic energy—like visible light but 10,000 times as powerful. The photon particle has no mass, no size, and no charge and wouldn't even exist without its energy. In fact, it stops existing when its energy is spent. Like all things subatomic, a photon functions as both a particle and a wave, and in a gamma ray that wave is vanishingly tiny, measuring less than a billionth of a meter. Because it is so powerful and travels so far, gamma radiation can be mapped from the air.

In contrast, even very energetic alpha particles don't travel more than a few inches in the air because they are composed of two protons and two neutrons and are thus quite heavy. Beta particles are much lighter, but even they travel no more than a few feet. This is why most maps of Chernobyl contamination in 1986 show only the cesium levels. Cesium emits a beta particle to produce barium-137, an extremely short-lived isotope that emits easily mapped gamma rays. But a brew of other radionuclides—atoms with radioactive nuclei—was also released.

Because the fourth reactor had been operating for two years, it was packed with an extremely radioactive inventory of fission products: cesium-134 and 137; iodine-131; strontium-89 and 90; plutonium-238, 239, 240, and 241; as well as a host of other radionuclides whose release from the core depended on their chemical qualities and the way the disaster developed over time.

Volatile elements (that is, those that easily turn to gas) such as iodine vaporized and bonded with mist, as did cesium, the only metal other than mercury to exist as a liquid at room temperature. The heat rising from the graphite fire's enormous temperatures created a smokestack effect that lifted the radioactive gases and dust a kilometer high into the atmosphere. The smokestack effect did much to spare the local population from the most lethal effects, but it also allowed the contamination to drift around the northern hemisphere, falling with rain to leave highly radioactive patches in Sweden, Bulgaria, and Austria and prompting the German Green Party to take up what became a global slogan: *Chernobyl ist Überall,* Chernobyl is everywhere.

Closer to the ground, the core sprayed radionuclides like a sprinkler, changing direction with the wind and contaminating patches of land around the reactor with shifting gradients that stain the radiation maps with deepening shades of mauve, rose, and brick.

When the wind shifted south again on April 30, the fire seemed smothered after helicopters dropped 5,000 tons of sand, clay, lead, and dolomite onto the core. The releases of radioactivity—while still about a Hiroshima a day—seemed to have stabilized.

Inexplicably on May 1, in a development that had never been predicted in any worst-case scenarios, the core began heating up and belching radiation that penetrated the thick cap of extinguishing materials. Again, volatile iodine and cesium vaporized first. But as the core got hotter—reaching temperatures of 3000°C (5400°F)—the fuel melted. Less volatile elements such as strontium, ruthenium, and zirconium vaporized and floated south towards Kiev just in time for the May Day parade—which took place as planned, children included—while officials said nothing about the danger and secretly evacuated their own families from the city.

At the time, of course, almost nobody knew this. But luckily, Soviet officials were far from the only sources of information. The clothes, hair, and skin of Western tourists who visited the affected areas on different days during the disaster gave Western scientists important clues about what was actually going in the reactor. The results of such atomic forensics later made it difficult for the Soviets to lie about the disaster. Although they kept things secret for as long as possible, eventually they did share data with their colleagues in Western nuclear industries, but opinions differ on whether any of them were fully honest with their respective publics.

People who visited Kiev before May 1 did not show evidence of radiation exposure, but those who visited on that date or afterwards did. Items dubbed the "Kiev trousers" and the "Minsk shoe" in the scientific literature carried radionuclides that could have been released only at extremely high temperatures and provided some of the first indications in the West that the core had melted and was spewing radioactive contamination over an extremely wide area inside the Soviet Union.

On the maps, the contamination pattern looks like a hand. The palm is painted in dark shades only within six miles of the plant, while the fingers, formed as the wind shifted slightly each day, fade gradually to pink, orange, and yellow 20 miles away. There were also the "anomalies": the pink, mauve, and rose patches hundreds of miles away from the plant where radioactivity came down with rain. A similar phenomenon was observed after atmospheric nuclear tests in the United States. Fallout from bombs detonated in Nevada in the 1950s came down with rain on Rochester and Albany in New York.

On May 5, the day after Orthodox Easter Sunday, the core melt was so drastic that the radiation release was almost as large as the first day. Exposure levels approached 1,000 roentgens an hour close to the reactor core, 60 roentgens at the neighboring reactor No. 3, 800 milliroentgens in Pripyat, and 20 milliroentgens in the town of Chornobyl. Though few people in Kiev knew the details, the magnitude of the disaster was starting to trickle down, and on May 5 the city's railroad and bus stations were packed with families trying to get their children out of the city.

Then, suddenly, the core melt stopped. No one is certain why, because no one knows why the core heated up in the first place. If it happened because fission had resumed, it may have stopped when the fuel melted into a magma of tendrils and blobs that flowed about the reactor's depths, physically separating the fissile materials enough to stop the chain reaction.

To this day, no one knows exactly how much of the core the disaster released and it may never be known. In 1986 the Soviets estimated that 3.5 percent of the core was released, including 100 percent of the radioactive noble gases, 20 percent of the radioactive iodine-131, and 13 percent of the cesium-137.

Although the force of the reactor explosion equaled less than one percent of the 13-kiloton atom bomb dropped on Hiroshima, Chernobyl released a much greater amount of radioactivity. If Hiroshima released 3 million curies, Chernobyl spewed anywhere from 50 million to 200 million, although the most accepted estimate is the lower one. About 75 percent settled in the European part of the Soviet Union. The remainder sprinkled across the globe. In northern Scandinavia, thousands of contaminated reindeer had to be slaughtered because reindeer eat lichen and lichen, which have no roots, very efficiently absorb chemicals and nutrients from the air—including the radioactive elements in the Chernobyl cloud. Small portions of radiation were even carried by birds, flying from their wintering grounds in Africa and arriving in Finland coated with fallout.

When clouds gathered over the Chernobyl region in May, bringing scattered thunderstorms, the Soviet military seeded them to induce rainfall farther from the area. The cloud seeding sparked rumors that the real reason for bringing contamination down over Belarus and remote parts of Russia was to keep the radioactive cloud from reaching Moscow.

Estimates vary about the exact amount of territory contaminated by Chernobyl in the most affected parts of the former Soviet Union. But whether by natural rain, cloud seeding, or the gentle deposition of dry fallout, it is safe to say that Chernobyl contaminated about 50,000 square miles with at least 40,000 becquerels of cesium per square yard (square meter). Close to the reactor, levels reached 1 million becquerels and in places like the torch monument many, many times more.

A becquerel represents one radioactive atom decaying per second. It is the wee cousin to the curie, a unit that represents the decay of 37 trillion radioactive atoms per second. If one curie is a large amount of radioactivity, one becquerel is very small. Although the becquerel is the preferred unit internationally, it is such a small amount of radioactivity that using it requires either writing a lot of zeros or, worse still, using scientific notation. To avoid either, I will generally use becquerel when referring to small amounts of radioactivity and curie to refer to large amounts.

The average human body contains 7,000 becquerels from the radioactive carbon, potassium, and other elements in a normal diet. Radioisotopes used in medical diagnosis contain 70 million (even that doesn't achieve the status of a "large amount of radioactivity" and amounts to about two thousandths of a curie). The average smoke detector contains 30,000 becquerels of radioactive americium, although its alpha radiation does not penetrate the device's plastic.

Imagine sprinkling the radioactive contents from slightly more than one smoke detector on each square yard of New York State. This approximates Chernobyl's environmental impact in the affected parts of Belarus, Russia, and Ukraine. If you then dumped the contents of 50 smoke detectors on every square yard of Rhode Island, you can imagine the impact on lands within about 20 miles of the reactor. Nevertheless, it is still hard to picture since most of us don't deliberately break open smoke detectors to see what 30,000 becquerels of americium looks like.

It is easier to imagine what happened on the grounds of the Chernobyl nuclear plant, which was blanketed with anywhere from 1,100 to 2,200 pounds of radioactively contaminated debris expelled from the reactor in the explosion.

EXODUS

The nuclear station and the towns of Pripyat and Chornobyl were islands of modernity in a sea of scattered villages, where the mechanized life of the Soviet collective farm still hadn't fully penetrated the thatched wooden cottages and folk traditions of the historically isolated Polissia region. Straddling the ethnic border between Belarus, Russia, and Ukraine, Polissia is Europe's largest wetland, a water-logged landscape of peat bogs, marshes, swamps, and forests largely contained within the Pripyat River basin. In the nineteenth and twentieth centuries, most of the wetlands were drained to expose organic peat soils that were fertile for a while, but their nutrients were soon exhausted and, without fertilizers, the land deteriorated. Good only for growing undemanding crops like potatoes and flax, Polissia had the worst land in famously fertile Ukraine and not particularly good land in less fertile Belarus. It was the least populated region of both countries, had the lowest levels of urbanization, and also had the lowest density of roads and rails. The primary industry was dairy farming, but even the cattle were not very healthy because the grasses in their pastures were not nutritious.

Electricity lit most of the unpainted wooden dwellings, but there were few telephones and almost no private cars. People grew their own food, made their own tools, wove their own clothes, and still honored their own pre-Christian gods and traditions. Stores sold the little that people needed to buy—bread, a little salt—neither of them iodized.

Even when radiation drifted invisibly into their lives in the days following the Chernobyl explosion, the Polissian peasants continued the ancient cycles of agriculture and husbandry, planting the fields, pasturing cows, tending the family orchards and vegetable gardens. As on farms everywhere, children too had their chores, which kept them outdoors. When news of the disaster began to spread, many families kept their children confined to their houses, but no one kept them from eating locally grown food. Weaning babies were especially vulnerable since most of their diet was fresh milk.

The main risk in the first postdisaster months was outside, where radioactive dust coated everything—trees, buildings, roads, animals, birds, clothing, trash, and meadows where domestic livestock grazed, as well as the gardens and cultivated fields that provided for local diets. Radioactive iodine coated the pastures and meadows grazed by cows and, within 48 hours of the initial explosion, it had already laced their milk. Within a few weeks, radioactive iodine was detected in milk that the American embassy in Moscow sent back to the United States for testing.

The air, too, buzzed with bombarding particles and gamma rays. Rural residents were especially vulnerable because their wooden houses offered less protection than the concrete high-rises of Pripyat. Had they not been evacuated, Pripyat residents who stayed in their homes would have received lower doses than their country cousins.

Radioactivity is, by its very nature, temporary. The unstable nuclei that Chernobyl dumped on the environment were destined eventually to decay, gradually reducing the amount of radioactivity outdoors, although each type of radionuclide has its individual way of decaying and does so in its own good time. Plutonium-239 has 94 protons, too many for its 145 neutrons to hold together. So it emits alpha particles to reduce the positive charges. Iodine-131 has too many neutrons, so it emits beta particles, which are twins of electrons, but born inside the nucleus rather than outside. Beta decay turns a neutron into a proton and is often accompanied by gamma rays. Because alpha and beta decays change the number of protons, both transmute nuclei into different elements. Plutonium-239 decays to uranium-235. Iodine-131 decays to the inert noble gas xenon.

Decay is triggered by a quantum event that is utterly random, so it is impossible to predict when a particular radioactive atom will decay. But the quantum world does observe the laws of statistics, making it entirely possible to predict the length of time needed for half of a large number of radionuclides to decay. This is known as the half-life. After 10 half-lives, any radioactive material will decay away to 0.1 percent of its original amount. After 30, the amount is too small to measure.

Like most radioactive isotopes, the fission products spewed out by Chernobyl were largely short-lived. Molybdenum-99, with a half-life of 66 hours, decayed away within about a month, as did neptunium-

239 and tellurium-132. But the corollary of a short half-life is high radioactivity because so many nuclei are decaying at the same time. So, for that month, the air was charged indeed, lobbing anyone on a contaminated patch with a subatomic barrage of ionizing artillery.

Alpha and beta particles and gamma rays are called ionizing radiation because they are energetic enough to knock electrons off atoms. This creates ions, which are charged up and more chemically reactive than they should be to avoid biochemical trouble. Yet while nothing but lead can stop a gamma ray, alpha and beta particles have to be pretty close to those atoms to do any ionizing. Ordinary clothes are a sufficient barrier against heavy alpha particles, which are stopped by a sheet of paper. Being much smaller than alpha particles, beta particles travel farther but are stopped by solid materials such as tin foil or the metal in cars. When the alpha and beta particles come from all surfaces, they are much more harmful. Skin exposure causes beta burn, with symptoms ranging from a kind of nuclear tan at their mildest to blisters, ulcers, and sores at the extremes. Many people who suffered acute radiation illness after the Chernobyl disaster also suffered enormous beta damage to their skin, a condition known as "cutaneous radiation syndrome." This can lead to radiation keratoses, overgrowths of horny layers of skin, or to disfiguring redness caused by dilated capillaries and arteries.

Of the ephemeral fission products, iodine-131 was the most dangerous. With a half-life of eight days, it was virtually gone in three months. But in addition to externally zapping living things with beta particles, iodine-131 was absorbed into the body, collecting in the thyroid gland as an ingredient of hormones. Taking iodine tablets before or immediately after a nuclear accident packs the thyroid so it doesn't absorb as much of the radioactive isotope, but the Soviet authorities distributed iodine tablets immediately only in Pripyat. In the surrounding villages, iodine prophylaxis was a week late, and in some irradiated locales, it didn't arrive at all. Compounding matters, the nutrient-poor soils that make Polissia a choice location for wormwood and mugwort have so little natural iodine that the region has a historically high incidence of endemic goiter. With no other supply of the critical nutrient, the thirsty local thyroids eagerly drank in the radioactive iodine-131, which then wreaked biological havoc by ionizing the atoms in the glands' cells.

The rural evacuations began one week after the explosion, right after the core began heating up again. Based on gamma radiation readings taken at various distances, three circles were drawn around the disaster area. The innermost circle extended about a mile around the No. 4 reactor. The second circle had a radius of 6 miles; the third circle, a radius of 18 miles. The last two distances correspond to 10 and 30 kilometers, respectively, and because "10-kilometer zone" and "30-kilometer zone" are official designations, I will use them throughout when referring to the evacuated zone.

The entire 30-kilometer zone was evacuated the first week of May, although there are many conflicting reports about the details. Soviet newspaper reports asserted that the people together with some 35,000 head of cattle and 9,000 pigs were evacuated, although the ultimate fate of the livestock is unclear. Some say they were slaughtered and their radioactive meat was diluted to safe limits by mixing it with clean meat in processed foods. There were similar unconfirmed stories about the 10,000 domestic dogs that were supposedly shot to prevent rabies.

By the end of the month, a 100-mile perimeter of barbed wire, guard posts, and watchtowers bordered the 30-kilometer zone, which was coming to be known simply as the "zone," in contrast with the 10-kilometer zone, nicknamed the *desiatka*, or the "ten." As more detailed radiation readings were drawn in subsequent months, larger areas were evacuated beyond the borders of the 30-kilometer zone in Belarus, Russia, and Ukraine. By the end of September, 116,000 people were uprooted from 188 towns and villages, including Pripyat and Chornobyl.

THE BARROWS OF BURAKIVKA

Their epitaphs take the form of signs—where each village's name is crossed out with a diagonal red line—that can be seen on roadways and thoroughfares throughout Eastern Europe, marking the spot where a place ends. Copies of the zone signs can be seen in Kiev's Chernobyl museum, while the originals still stand where they always did, at the ends of what are now ghost towns and villages. Faded and scratched after 18 years, the most moving signs mark "no-places," fields of rugged hillocks that jut crazily like large toppled blocks blanketed with grass. These are the buried villages in the 10-kilometer zone like Yaniv, which sat directly in the western arrow of lethal radioactivity, less than two miles from the plant.

Enlarged and grainy photographs in Kiev's Chernobyl museum document how bulldozers interred Yaniv's traditional thatched-roof cottages: tipping them into pits, covering them with soil, and then flattening them into a desert while water trucks sprayed a steady rain to keep the radioactive dust down.

Many radioactive graveyards were created on the run, as portrayed in this affecting description in a Soviet newspaper: " . . . machines kill machines. A metal 'rhinoceros' occasionally visits a bus and, in a flash, instead of a bus there is a flat cake. A herd of vehicles dies before our eyes. And the 'rhinoceros' having dug a shallow hole, shoves all the flat cakes, unhurriedly tramples on them, getting the radioactive garbage out of sight as quickly as possible."

In the course of the cleanup, 600,000 "liquidators"—military and civilian—were sent to the zone on 15-day tours of duty to strip the surfaces of contamination. Not all surfaces. The zone was more than a thousand square miles and would grow with time. The Soviet budget was not large enough to clean it all, so decontamination in Ukraine was limited mainly to roadways and shoulders, the nuclear station, Pripyat, and the town of Chornobyl, which was just outside the 10-kilometer zone and became headquarters for the recovery efforts. The town's cleanup included razing a dozen "hot" buildings, removing 150,000 cubic meters of radioactive soil, and laying 10 miles of fresh asphalt and concrete on roads and pavement.

With the exception of a few villages and the town of Bragin, the contaminated parts of Belarus were largely left alone. The same was true in the Russian republic.

The Chernobyl cleanup in the Ukrainian republic almost matched the disaster in its magnitude. Buildings were blasted with sand, then washed and sprayed with liquid glass to fix whatever radioactivity was left. Nearly 11 miles of dikes and dams were built to keep radionuclides from spilling into the Pripyat River, a major tributary of the Dnieper and source of much of then-Soviet Ukraine's water supply. Roadsides were completely stripped and buried, and the roads themselves were repaved. Five thousand hectares of surfaces were sprayed with chemicals to keep radioactive dust from rising in the hot and dry summer of 1986. The roads in cities such as Kiev were washed so frequently that the vegetation was lush despite the lack of rain.

To keep radioactive dust from spreading outside the zone, the thousands of vehicles and machinery brought in for the cleanup were destined never to leave. Liquidators working their two-week tours of duty changed into "dirty" buses at the 30-kilometer zone checkpoint and, if they were working in the "ten," changed into still more contaminated buses at the inner checkpoint. They changed buses again upon leaving each zone.

Cleaning up the nuclear station's grounds proved the greatest challenge. Moscow had no intention of abandoning the Chernobyl plant, which accounted for 15 percent of the Soviet nuclear energy capacity and more than 80 percent of its energy exports, mainly to Hungary. Indeed, the Soviet government's insistence on continuing to operate the plant was one of the issues that fueled Ukraine's drive for independence in 1991. Once independent Russia presented independent Ukraine with its energy bills for fossil fuels, however, operating Chernobyl didn't seem so bad to Kiev, which kept the plant running until pressured by the West to close it in 2000.

But back in 1986, there were different priorities. Highly radioactive debris on the neighboring No. 3 reactor's roof had to be shoveled off by hand. The fuel fragments and graphite littering the station grounds were bulldozed together with concrete, asphalt, and 20-foot-deep layers of topsoil. They were then stuffed into a wall of the 20-story concrete and steel structure built around the ruined reactor to contain the invisible clouds measuring hundreds of roentgens that hovered over the crater. It took 90,000 liquidators nine months to build the protective shell, officially called the Shelter Object, though better known by its nickname, the Sarcophagus.

The completion of the Sarcophagus in November 1986 ended the first stage of the Chernobyl recovery, and reactors No. 1 and No. 2 were restarted. By then winter had set in and the 10-kilometer zone was a barren wasteland, as desolate and bleak a place as any imagined in the most apocalyptic science fiction. It seemed as though a part of the planet had been killed (Plate 2).

Valery Antropov worked for "Complex," a specialized government agency for radioactive waste management and decontamination. I met up with him at Burakivka, the site of a World War II mass grave of Soviet soldiers on the very tip of the western arrow, six miles away from the reactor. After Chernobyl it became a grave for short-lived radioactive waste, enclosed in barbed wire. While Rimma, the guide, and Svetlana, the botanist, drank tea inside the facility's cinder-block headquarters, Antropov and I strolled amid the rusting metal carcasses of bulldozers and excavators, cranes and compactors, dumpsters, loaders, and the armored personnel carriers whose layers of lead shielded their occupants from gamma rays in the early postdisaster months. The equipment had been brought from all over the Soviet Union to work on the cleanup and to be left in the zone for good, too contaminated to ever leave.

"What are we supposed to do with all this?" Antropov asked rhetorically, waving his arm at the jumbled equipment and machinery. There were three such equipment graveyards in the zone. "We can decontaminate metals by putting them in large pools of acid that remove the outer radioactive layers. That works well for smooth objects like pipes. But machinery is more difficult because its surfaces aren't smooth."

With too many nooks and crannies for radionuclides to hide in, the machinery should be dismantled and safely buried. But of the 30 long-term storage trenches at Burakivka, 26 have already been filled with 10 million tons of waste. The remaining four are far from enough for the additional 10 million tons that already require long-term storage, with much more still to come.

Antropov's agency was also in charge of the protective gear that I rented from Chernobylinterinform for $15. Army surplus left over from the days when the Soviet military spearheaded much of the Chernobyl cleanup, it is convenient outerwear and the ubiquitous zone fashion. Although it offers no more radiation protection than regular clothes, if I did get some funky dust on me, I could just give it back to Antropov for decontamination and get another pair. But it was made of blue vinyl, swished annoyingly, and made me look three times my actual size.

It was but a few minutes' walk from the radioactive parking lot to the burial mounds. The skies had grown steely gray, but there was no wind and the only sound was from the gravel crunching beneath our feet as Antropov showed me around the 250-acre site. Burakivka is one of the most radioactive places in the 30-kilometer zone—though places inside the "ten," and especially near the Sarcophagus, are still higher.

The mounds resemble the huge prehistoric burial mounds called *kurgans* that dot the Ukrainian steppe. The kurgans' most ancient burials are more than 5,000 years old, but the barrows of Burakivka need last only 300 years. By then the radioactivity will have decayed away.

"This is the largest storage facility for nuclear waste in the world," Antropov said with pride. To a first-time visitor, such enthusiasm might have seemed odd, even bizarre. But it is quite common in the zone. The dedication and friendliness of the people who work there make it a surprisingly pleasant place to visit.

We came to a sandy bank overlooking an unfilled trench that was marked on the Burakivka map Antropov had given me as a gray rectangle—No. 28. About the size of an Olympic-sized pool and 12 feet deep, the bottom was a shallow marsh of reeds and sedges, their dried seed heads ready to scatter come spring. Blanketed on its floor and sides with a four-foot layer of clay, the trench was impermeable to water and the radionuclides that can flush out with it. The marsh formed from rain with no place to drain, though it evaporated in dry seasons.

"The Soviets used clay because it was cheaper than concrete, but it turns out that water can eventually penetrate concrete but not clay. There's been no leakage at all in 15 years," explained Antropov, leading me past trench No. 27, which was stuffed with a high pile of exposed debris.

Tractors would eventually compact the junk and cover it with a thick external blanket of clay. Once the debris settled even more inside the protective cocoon, the mounds would be packed with soil and planted with grass to keep them from blowing away. The lawns must be carefully tended and weeded of any stray saplings, because tree roots would crack the clay coffins, much as they are cracking through towns like Pripyat.

The complete sites are colored green on the map, an oddly ecological color considering their contents. But the point of managing radioactive waste is not to eliminate but to isolate. There is no currently practical way of getting rid of radioactive waste, a sobering fact that fuels much antinuclear sentiment. It can only be stored in such a way that radionuclides do not get into the environment. The barrows of Burakivka are green because they are among the few places in the zone where the millions of tons of radioactive Chernobyl debris are stored safely.

"Points for the temporary localization of radioactive wastes" is the cumbersome official term for Antropov's biggest professional headache. It is a fancy way of saying "leaky radioactive dumps."

"Digging, dumping, and covering were the right thing to do in

1986 so that all of the radioactivity wasn't on the surface. It reduced background radiation by many orders of magnitude," Antropov told me when we were back in the car with Rimma and Svetlana, driving towards the town of Chornobyl. "Those were emergency methods. They were only supposed to be temporary. But the dumps are still in the same places after 15 years."

One problem is that all of the dumps are leaking into the environment. Without clay seals like those in Burakivka, water trickles through the debris and washes radionuclides into the soil.

Antropov pronounced "points for the temporary localization of radioactive wastes" as a single word based on its Russian acronym PVLRV: *peverelev*. Though he understood my Ukrainian perfectly well, Antropov, like most people in the zone, usually conversed in Russian. The specialists brought in during the early postdisaster years, when the zone was controlled directly by Moscow, came from throughout the Soviet Union and the lingua franca was Russian. It remained so even after the Ukrainian and Byelorussian (or Belarusian) Soviet Socialist Republics took legal control over the parts of the zone on their respective territories in early 1991. Ukraine got the larger part of the 30-kilometer zone, together with the nuclear station and waste dumps, and called it *Zona Vidchuzhennya*, a name it retained when the USSR collapsed in the summer of that year. Most Ukrainian signs translate the name into English as "Exclusion Zone," though a better translation for *Vidchuzhennya* is "alienation." I use both, but find Zone of Alienation a more affecting and accurate name.

Another problem with the waste dumps is that no one knows where they all are.

"We used to say that there are 800 *peverelev*," he said. "But now we've come to realize that we really don't know how many there are," he continued as we drove past Kopachi, another buried village about midway between the two Chernobyls—the station and the town. On the radiation maps, it is the dark red color of brick.

He pointed out the window at the rough-hewn hillocks that bordered the empty road. We hadn't seen any other cars in more than three hours of driving around the 10-kilometer zone.

"See, there you can tell that something is buried," Antropov said. "But many of the burial trenches have flattened with time and not all of them were marked or mapped, so we don't know how to find them."

Svitlana, who was in the back seat with Antropov and me, also looked out the window and jerked her chin at what looked like a shrub on one of the hillocks.

"You can find them by the radiomorphism of the trees," she said, explaining that plants change their shape under the influence of high radioactivity. Instead of growing like a tree, which has a large, single trunk, and instead of growing like a Scots pine, on which the branches grow perpendicular to the trunk and well above ground level, radiomorphic pines grow from a single trunk close to the ground but branch into a filigree of multidirectional stems. They look like pine bushes.

"In 1986 and 1987, many trees exhibited radiomorphism. But today, the only places where radioactivity levels are high enough are the dumps," said Svitlana. "That's how you can tell where they are."

Clearly intrigued, Antropov asked: "So, if we did a helicopter survey, we could identify the *peverelev* by the trees?"

Svitlana nodded: "The trees grow like bushes so they will be shorter than the surrounding forests."

The zone was an interdisciplinary problem. The two specialists were still talking about identifying radioactive waste dumps with stunted pine bushes when we drove up to the 10-kilometer zone checkpoint. The first time I visited Chernobyl in 1996, I had to change out of a "dirty" 10-kilometer-zone car at that very same spot, although it had no longer been necessary to change cars at the border of the 30-kilometer zone to go back to Kiev.

Five years later, there were still checkpoints on the borders of the 10- and 30-kilometer zones. But I could drive my own car from Kiev into the "ten" and back again, and I usually did drive my own car in researching this book. It was cheaper than hiring drivers or taxis—the only other alternatives since public transportation doesn't go to any of the evacuated and resettled regions.

That I was even permitted to do so was because a great deal of radioactive isotopes have already decayed away. After the extremely short-lived radionuclides disappeared in the first postdisaster year, the main external radiation doses came from cerium-144 and ruthenium-106, with half-lives of approximately one year, and after 15 years, only tiny fractions of them were left. At the dawn of the third millennium, the main radiation danger came from cesium-137 and strontium-90. Plutonium, too, will be a problem for all of imaginable time. But 95 percent of the radionuclides that remain are no longer *on* the zone, in the form of dust and fallout that could get on me or in me. They are now *in* the zone, sunken to a depth of about two inches in the soil whence they have insinuated into the food chain.

"Chernobyl" and all that word entails is no longer a state of shock but has become a state of being—a radioactive state never encountered in nature on such a scale before.

NIGHTLIFE IN CHORNOBYL

By law the home base of any visitor to the Ukrainian Zone of Alienation (Belarus has different rules) is in a daffodil yellow prefab building in the town of Chornobyl. Brought from Finland in 1986 to house the Soviet emergency committee overseeing the cleanup, it is now the headquarters of Chernobylinterinform, the government agency that ushers nearly all visitors to the zone for the duration of their stay, except for evacuees and their families returning for visits. Much like the old Soviet Union, the zone is not a place in which to let people wander around freely. It is a land of internal checkpoints and watchtowers. Chernobylinterinform is like a neo-Intourist, the official Soviet travel agency that exercised total control over all visitors. It even has a hotel, in another yellow prefab building across the road, where I stayed during my visits.

But like the watchtowers, which are for fire prevention, Chernobylinterinform's control is largely benign, to ensure a visitor's safety rather than a paranoid regime's secrecy—which was certainly the case in the early postdisaster years. You need permission to enter the zone, but after 15 years, getting it is routine. Just send a fax to Chernobylinterinform, tell them what you want to do, and give approximate dates. After that, you will get a return fax with a proposed program. According to zone regulations, the program is the equivalent of a permit to be in the zone.

After many zone visits, I've found the agency's staff remarkably accommodating. Maryna Poliakova oversees the process. A matronly blond with a sad smile and stylish shoes, she works Mondays through Thursdays in a cozy office that is always the first stop for the hundreds of delegations—Swiss scientists, Japanese tourists, Ukrainian journalists—that visit the zone annually.

After a contraption that looked like a futuristic Nautilus machine flashed a green light to assure me that there was no radioactive dust on my hands or feet, even after traipsing in the wormwood fields and forests all day long, I went to Maryna's office to go through my notes while waiting for an early dinner in the canteen down the hall. Radiation maps and colorful calendars decorated the walls.

Rimma came in with a cloudy glass bottle and three shot glasses. "Let's get warmed up," she said, pouring viscous liquid into the glasses and passing them around. The distinct odor of homemade vodka filled the room.

Maryna sniffed her glass. "Is this from that *samosel?*"

Samosels means "squatters" or, more literally, "self-settlers." The term refers to the several hundred people, overwhelmingly older women, who returned after the evacuation to live semilegally in the zone.

"Is it radioactive?" I asked, feeling slightly foolish. Rimma has a wealth of stories about foreign journalists who come to the zone in surgical masks and fear every puddle of water as a potential fountain of radioactive waste.

"No, it's made from sugar, and radionuclides doesn't get inside sugar beets," said Rimma. "Besides, the sugar is from outside the zone."

In fact, as I was told by scientists later, radionuclides *do* get inside the sugar beets, but they don't remain in the processed sugar. For the purposes of deciding to drink or not to drink, though, it was a distinction without a difference.

Rimma nodded at me to try it. "It's great stuff! I got it from an old *babushka*," she explained, using the affectionate nickname for a grandmother.

Actually, it was foul beyond words. I grimaced after taking a sip and gave Rimma a puzzled look.

"Sixty percent alcohol," she said. "You'll see. It grows on you."

She was right. It was almost as potent as absinthe, and probably as bitter. But whether it deadened my taste buds or had some other mysterious effect, after a second shot I did grow to like the stuff and acquired a pleasant glow just in time for dinner.

After a hearty four-course meal, I took a stroll outside. Chornobyl is not a ghost town like Pripyat. It is where the administration of the Zone of Alienation performs its dystopian task of running the no-man's-land. It is also where zone workers like Rimma and Maryna live during their tours of duty. Unlike Maryna who works four days a week, Rimma works two weeks and then has two weeks off when her replacement takes over. Zone rules prohibit anyone from living or working there full-time, which is why the *samosels* are in violation of the rules.

A labyrinth of silvery pipes snaked around the streets. Like most provincial Soviet towns in 1986, Chornobyl had old pipes that couldn't deliver enough hot water for the two daily showers liquidators had to take during the cleanup. But new pipes couldn't be laid in the ground because it would stir up radioactive dust. So they were laid on the surface, where they remain to this day, forming square arches where they cross the roads, like bridges.

Aside from the eponymous plant seven miles away, where about 4,000 people work, the town of Chornobyl is the most populous place in the no-man's-land. But "populous" in this case means about 2,500 people daily in a town that once housed 10 times that number. Only a few curtained windows in the low-slung apartment buildings bespoke of any human habitation.

After about 10 minutes the asphalt road turned into a rough path, passing empty cottages, their gardens and orchards overgrown with weeds, bushes, and birch trees. The encroaching forests have consumed much of the wood-and-plaster cottages, even more so than the concrete and cinder blocks of Pripyat. Massive tangles of wild grapes crush thatched roofs with their weight. Trees shatter walls with the force of their growing branches and smash through buildings completely when they fall. Microbes and fungi feast on the organic materials in the wood, resins, paint, and paper used in building interiors.

It seems odd, but it is impossible to smell fresher air in an inhabited urban setting than in Chornobyl, where the number of cars can usually be counted on one hand and songbirds frequently provide the only sound. It is one of the disaster's paradoxes, but the zone's evacuation put an end to industrialization, deforestation, cultivation, and other human intrusions, making it one of Ukraine's environmentally cleanest regions—except for the radioactivity.

But animals don't have dosimeters. With nearly all humans in the zone confined to the two Chernobyls, the rest of the Rhode Island-sized territory has become a fascinating and at times beautiful wilderness teeming with beavers and wolves, deer and lynx, as well as rare birds such as black storks and azure tits.

At waist level, my dosimeter displayed a reading of 40 microroentgens per hour on the path, but when I stepped off it into the brush growing near a collapsing cottage draped in scarlet leaves, the reading shot up to 70 and still higher when I crouched to hold the instrument closer to the fallen foliage. Then I noticed that the cottage's yard had been stripped and the ground chaotically plowed by boars. The zone's most populous large animals, boars roam its vast expanses in large herds. Nearly everyone in the zone has a story about encountering the woolly, tusked pigs in Pripyat courtyards, village gardens, and the town of Chornobyl, which the boars like to visit in the autumn for the windfall fruits in abandoned orchards.

Turning back to return to the hotel, I approached a camouflage-clad woman with short blond hair and a gold molar that flashed when she chortled at me.

"So, you're human!" she said, pleasantly enough. "I saw something dark in the bushes and thought it was a boar!"

I laughed, too, at being mistaken for a boar, and after joining me in a few chuckles, the woman went on her way.

A blackbird trilled a song in the distance and then suddenly stopped when a golden eagle soared above some nearby tree tops, hunting for an evening snack. Golden eagles are rare in the forests of northern Ukraine, Belarus, and Russia; it was very rare to see one flying over a town, even if it was a mostly abandoned town like Chornobyl. It was also very rare to see one so close. Even a highly inexpert birder like me was able to identify it when I checked my bird handbook later. But the feral fields surrounding the largely empty town are now rich with the hares, rabbits, and rodents that are the golden eagle's favorite diet. The multitude of small creatures has attracted many raptors, and it is almost impossible to visit the zone without spotting one, at least at a distance, hovering on a thermal.

The street around the hotel was devoid of people, and twilight started falling as I approached the yellow buildings. A small patch of lawn behind Chernobylinterinform had also been rooted by boars, although the light covering of fallen leaves suggested this had occurred some time earlier. No wonder boars are the bane of farmers, ripping up cultivated fields and causing much economic damage.

But there are no farmers for a boar to bother in the zone, I thought, climbing the short flight of stairs to the sparsely furnished but comfortable suite that served as my zone accommodations. There was even cable TV. During busy seasons, usually around the disaster's anniversary, Chernobylinterinform's hotel can be a happening place. But October is slow, and the corridor was so dark I needed my penlight to check the room number on my door.

Taking off my camouflage outerwear, I dropped it in a separate pile on the floor, far from the couch where I settled to finish my notes. Reason and the flashing green light on the radiation detector told me they were uncontaminated. My dosimeter beeped a perfectly normal reading of 12 microroentgens an hour.

Yet the dread of radiation lingered. I recalled my exasperated fifth-grade teacher warning our misbehaving class that we would die in a nuclear war because we didn't follow instructions. I spent months afterwards avoiding the radio for fear of hearing the Emergency Broadcast System announce: "This is NOT a test." Inexplicably, that fear did not extend to television.

But the twin terrors of nuclear apocalypse and radioactive desolation have been constant companions of the nuclear age. Life, if it managed to survive the holocaust, would be horribly mutated, like the monsters inhabiting the contaminated badlands of science fiction. But Chernobyl was showing me a different view of the future. It was a radioactive future, indeed, in which ghost towns and villages stand in tragic testimony to the devastating effects of technology gone awry. But life in the Wormwood Forest was not only persevering, it was flourishing.

2

Four Seasons

There will come soft rains and the smell of the ground,
And swallows circling with their shimmering sound;

And frogs in the pools singing at night,
And wild plum trees in tremulous white;

Robins will wear their feathery fire,
Whistling their whims on a low fence-wire;

And not one will know of the war, not one
Will care at last when it is done.

Not one would mind, neither bird nor tree,
If mankind perished utterly;

And Spring herself, when she woke at dawn
Would scarcely know that we were gone.

Sara Teasdale

One of the few patches of earth that Chernobyl radiation made entirely uninhabitable was, ironically, one of the first parts of Europe to be inhabited by modern humans. When continental ice sheets blanketed much of Belarus during the last Ice Age, the zone was part of a periglacial steppe. It was a forbidding, arid tundra periodically plagued by blinding dust storms of silt called loess that blew in from the glaciers. But about 25,000 to 30,000 years ago, small bands of hunter-gatherers made their way north from Africa. Equipped with fire, warm clothes, and sturdy mammoth-bone dwellings, they

were able to brave the daunting conditions for the superb hunting op-
portunities offered by large grazing herds of mammoth, bison, and
ancient horses.

People stayed on after the glaciers retreated 12,000 years ago and
tundra steppes gave way to forests and swamps. As human populations
grew and Ice Age game such as mammoth disappeared, new generations of hunter-gatherers aimed for nimbler quarry like deer but turned
more and more to fishing and gathering the fruits of the forest for
much of their food. When the Neolithic revolution brought pottery
from the Near East to north of the Black Sea around 5500 B.C.E., the
denizens of what is now Polissia impressed their pots with comb and
pit symbols that made them part of a larger archaeological identity
known as the "Comb-and-Pit" culture.

There is little evidence that the forest dwellers moved very much
over the mute millennia that followed, although combs and pits were
followed by a succession of different pottery, tools, and weapons that
hint at the cultural morphing and tribal mixing that culminated in the
first written reference to the Comb-and-Pit folks' descendants. In the
fifth century B.C.E., the Greek "father of history" Herodotus wrote of a
people he called the Neuri, who lived in the region before being forced
to abandon it by an invasion of snakes.

The Neuri were most likely Iron Age Balts, and the snakes that
chased them out may have symbolized Slavs. In any case, the next written mention of the forest denizens identified them as a Slavic tribe
called *Drevlyany,* or *Derevlians.* With a name derived from *derevo* (or
drevo in the ancient form), which means "tree," the woodland Slavs
were one of the founding tribes of Kievan Rus'—the medieval state to
which Belarusians, Russians, and Ukrainians trace their roots. Indeed,
the place name "Chornobyl" first appeared in an 1193 charter that described a Kievan Rus' prince's hunting lodge.

In the centuries after Kievan Rus' declined, the *Drevlyany* lands
passed politically from the Grand Duchy of Lithuania to Poland, from
Poland to Russia, and from Russia to the Union of Soviet Socialist Republics (with a brief interregnum in the short-lived independent
Ukrainian republic); then, after the USSR's collapse in 1991, they were
split between independent Belarus and Ukraine. But the people who
lived there—like more than half of Ukraine's population—remained
largely of the same genetic stock as the original Ice Age mammoth
hunters. Their evacuation in Chernobyl's wake brought an abrupt end

to some 25,000 years of continuous—if not uninterrupted—settlement in one of humanity's most ancient European homelands. Yet the sad cultural absence of people has allowed nature to resume its primordial patterns and cycles. Only now, radiation has become their integral part.

WINTER FALLOUT

If all you have is a cheap handheld dosimeter, you may as well keep it in your pocket when strolling in the Red Forest. I learned this rather quickly when mine simply shut down from the overload of being in multiple milli-land.

Rimma Kyselytsia, the Chernobylinterinform guide, and Svitlana Bidna, the botanist, meandered with me on a freezing, gloomy day in a forest of pygmy pines, twisted and bushy like the trees that grow on leaky nuclear waste dumps and then some. In fact, some plants resembled neither bush nor tree.

Svitlana stopped by one pine that had started out normally, with more or less perpendicular branches, but then sprouted what resembled a large upright broom from the top. Another pine was twisted into filigree (Plate 3).

I wished that we had a more powerful dosimeter to let us know what the exact levels were. But there were few places in the Red Forest with levels lower than five milliroentgens an hour. This meant that if we stood in such places for six hours, we would be exposed to the equivalent of a chest X ray (30 milliroentgens).

Looked at another way, the maximum dose considered safe for people who work in the nuclear industry is 5,000 millirem—which is the same as five rem—a year. For civilians, the limit was a tenth of a rem a year. We'll define "rem," "roentgen," and other radiation units later, but for now we can consider rem roughly equivalent to roentgens. Although I wasn't sure if I had entered the category of quasi-nuclear worker by writing a book about Chernobyl, I knew that the Red Forest was not a place in which I wanted to spend too much time.

Actually, the name "Red Forest" is a misnomer because it isn't red and, with its short and stunted pines, it isn't much of a forest either. But in 1986 the four and a half square miles of evergreen woodlands stood directly in the path of the deadliest debris from the explosion and then turned red before they died.

Those trees and about 1 million square meters of topsoil were buried on the spot in a "point for the temporary storage of radioactive waste" and covered with four feet of sand. The sand was sprayed with a liquid polymer that hardened to keep it from blowing away. After 15 years, there were still patches of polymer here and there, like flat sandstone pancakes that are easy to spot because nothing grows on them. Indeed, had the polymers not cracked and disintegrated, parts of the zone would resemble a barren moonscape to this day. But the sand was eventually planted with the young pines that surrounded us. Although the pines—regardless of their strange shapes—are green, the nickname Red Forest stuck and it now refers to one of the most radioactive outdoor environments on the planet.

The trees themselves are very radioactive, too, containing up to 500,000 becquerels of cesium and 7 million becquerels of strontium in a kilogram of wood!

Rimma flicked her chin at the surrounding expanse of stunted trees. "There are places in the Red Forest where background is one roentgen an hour."

The last sentence was much easier to write than to experience. It is one thing to sit in a safe office and write about what radionuclides do in the wild. It is quite another thing to stand in their midst while guides toss around alarming figures. One roentgen was a lot. Though I try to affect nonchalance when the gauges start inching into the milli-range, and usually succeed well enough to avoid the patronizing smiles that excessive radiation fears can prompt in zone professionals, my concern was obvious.

Rimma smiled, but without patronizing me. "Don't worry," she started to say before I interrupted: "You always say that."

"I say it when it's true. Those high levels are at isolated points, hot spots," she explained. "We know where they are and we're not taking you there. Besides, one roentgen measures the radiation you'll be exposed to in an hour. It doesn't mean that you'll get a dose of one rem."

"I knew that," I responded, just a split second too quickly.

Radiation exposure, measured in roentgens per hour, can be likened to a battlefield where the air is streaking silently with invisible particles and rays. The fiercer the exchange of artillery, the higher the exposure will be. Even if you were to stand in the midst of battle, though, not every rocket or bullet would hit you. Those that do make up your absorbed dose are measured in units called "rads." But in the same way that being nicked on your arm by a bullet differs from being shot in a vital organ, radiation has different biological effects—even for the same amount of absorbed dose—depending on the type of tissue it hits and whether the radiation is alpha, beta, or gamma. Biological impact is measured in rem units. For gamma rays and X rays, the rem dose is the same as the roentgen exposure. Dose is the same as exposure for beta particles as well. But for heavy and energetic alpha particles, the rem dose is 20 times the exposure level.

Based on morbidity and mortality rates after the atomic bombs were dropped on Hiroshima and Nagasaki, an acute dose of 100 rem is the minimum needed to trigger acute radiation illness. Acute doses of 300 to 500 rem will kill most people, although new treatments have increased survival rates. Acute doses of more than 1,500 rem will kill anyone.

From this perspective, one roentgen didn't seem so bad—especially since nearly all of the radiation exposure you get in Chernobyl these days is chronic, not acute. Besides, it was far away. Like staying off the battlefield, the greater your distance from a radiation source, the less likely it is to lob you with its atomic artillery.

There, however, the battlefield analogy breaks down. If the atomic world behaved anything like the classical physical world in which we live and where real battles take place, radiation would be impossible. In the classical world we experience with our senses, a bullet can put a hole in a wall and emerge from the other side. No matter how many times that bullet is shot, however, it will never pop through the wall without leaving a trace and suddenly appear outside it.

But that is exactly how an atomic nucleus emits radiation. In fact, it only appears to be "emitting" to us, as we imagine, say, an alpha particle shooting through the nuclear boundary like a bullet. But this is impossible. No alpha particle is energetic enough to escape the strong force. What it can do, however, is some quantum magic because it has a dual personality as a wave. It isn't a physical wave like a gamma ray. It has no energy and can't be detected because if it is detected, it immediately stops being a wave and becomes a particle.

The quantum wave reflects the weird subatomic reality that a particle such as an alpha becomes an alpha—that is, an identifiable thing with dynamic attributes like speed and position—only when it is measured. Now, measuring a quantum particle is in and of itself a paradoxical, controversial, and poorly understood process at the interface of mind and matter, which seems to require both consciousness and ignorance. But before that act of measurement takes place, the alpha exists in the abstract form of a mathematical formula known as a wave function. Symbolized by the Greek letter ψ, or psi, the wave function represents the statistical probability of finding that alpha in a particular place at a particular time. Before they are measured, the probabilities are smeared all over the place. Measuring psi focuses those fuzzy probabilities into a specific place and time.

Because of the fundamental quantum uncertainty about where—and whether—anything really "is," the subatomic occupants of a nucleus are less like bullets in a gun and more like the water in a washing machine. The particles' probabilities spin and cascade around inside, sloshing and splashing against the energy barriers imposed by the strong force. If the nucleus is filled normally, as in a stable isotope, the probabilities will always splash back inside. Although the universe is mind-bogglingly large and full of wonders for which our imaginations are inadequate, a stable atomic nucleus has never been known to decay.

However, if it is overfilled, like a radioactive nucleus, some of the probability eventually leaks outside the energy barrier. If the barrier is small and the leaking ψ is large enough to penetrate it, the probability of finding the particle outside the nucleus becomes not zero and, therefore, not impossible. In the quantum world, anything that is not impossible sometimes happens. So, if a droplet of psi leaks out of the barrier, it can pull the rest of the probability wave with it, making it appear that a particle has "tunneled" out of a nucleus without leaving any physical trace that it had passed through.

The same principle fuels fission, except in the reverse, when a neutron tunnels into a nucleus. That's why slow neutrons are better at fission than fast ones: neutrons that move slowly are in the vicinity of a nucleus longer, giving their probability wave more time to leak inside.

In quantum terms, the Red Forest was leaking a lot of probabilities. As we trudged through the packed sand among the twisted pines, I wondered if there was any deeper significance to the fact that ψ looks much like a stylized trident, the symbol of the Kievan Rus' princes that the first independent Ukraine adopted as its state seal in 1918, followed by the second independent Ukraine in 1991 (Figure 2). In fact, the classic trefoil symbol of radiation was also a tri-sign.

Svitlana motioned for us to gather around a straggly pine. With a

988   1918   2001

FIGURE 2 Tridents over the ages.

gloved hand, she fingered a branch that was bare but for a few clumps of stunted, tightly bunched needles that resembled thistles.

"Normal pine needles are supposed to be about an inch long and grow in pairs on either side of the branch. Radiation damages their spatial orientation so the shoots grow in the wrong direction and in the wrong places," she explained, pointing out another branch that had sprouted curly minibrooms. "In the early days, there was a lot of radiomorphism. Now you just see it in places like the waste dumps and here in the Red Forest."

Since the dawning of the nuclear age—especially in the late 1950s and 1960s when there was growing concern about the effects of fallout from atmospheric nuclear testing—much has been learned about what acute external radiation does to plants at places such as the Savannah River site in the U.S. Department of Energy's nuclear weapons complex. But nothing in the outdoors of any DOE facility compares to the levels and extent of radiation in the Exclusion Zone, where scientists (including scientists from Savannah River) can comprehensively study the effects of chronic and constant radiation exposure on vegetation in the wild, with many surprising conclusions.

As we shall see, Chernobyl's radionuclides behaved very differently in the wild than anyone expected. But initially, they did what any kind of fallout will do: they stuck to things (Plate 4). The end of April is early in the growing season for Polissia. In 1986 the collective farms' fields had already been sown and leafy trees were just beginning to bud, so it fell to the conifers to intercept as much as 80 percent of the radioactive cesium that coated the 30-kilometer zone. Fallout stuck especially strongly to wrinkled leaves, where hot particles could embed in the creases, and also on sticky surfaces like those on buds, leaves, needles, and certain flowers. Given their sticky resin and gnarly bark, pines were particularly magnetic.

With fallout on just about everything in the surrounding environment, the plants' radiation doses were partly external, primarily from gamma rays. But their greatest doses came from the radionuclides stuck to their own surfaces and bombarding them not only with gamma rays but also with alpha and beta particles, whose biological damage capability increases at tiny distances.

Because radiation does the most damage to cells that are actively dividing, growing organisms are most vulnerable to its impact. In mammals, active cell division and growth slow significantly in adults except for cells in the hair, skin, bone marrow, and gastrointestinal tract, which is why high radiation doses cause, among other symptoms, hair loss and vomiting. But a plant continues growing throughout its lifetime, and its most active cells are in the perpetually young tissues called meristems. Located in buds, root tips, and the cambium in the outer layers of stems, branches, and trunks, meristems are like botanical stem cells. With the exception of roots, which were hidden in the ground, these were precisely the plant parts to which radiation adhered, so what did grow in 1986 grew more slowly and much more strangely than usual.

The specific effects depended on the plant species. Radiation resistance among trees, for example, varies widely. Evergreens die at lower doses than leafy trees and young trees are more vulnerable than older ones. The growth of spruce and pine, the most sensitive species, gets stunted at absorbed doses of 150 to 250 rads. But it takes 1,000 to 1,500 rads—fatal for most other trees—to slow down an aspen. Birch and alder are somewhere in the middle.

Inside the 10-kilometer zone, where radiation levels were highest and the doses were close to lethal, plants' organs changed their shape or size. Leaves and flowers were wrinkled or twisted, or they grew asymmetrically. Irradiation of buds caused smaller, bushier, and more asymmetrical shoots. Norway spruces, oaks, lindens, and horse chestnuts displayed gigantism, sprouting leaves and needles that were much larger than average though their shape was perfectly normal.

In places where gamma radiation was about 30 milliroentgens an hour, field sagewort buds sprouted short shoots, thickly covered with clusters of deformed leaves, while purple loosestrife grew normally at first, but the tips were thickly covered with narrow leaves instead of flowers.

Although a 30-milliroentgens hourly exposure is far from fatal, a pine needle growing under such a background level can absorb a total dose as high as 500 rads over its two-year lifetime.

Such a prolonged dose doesn't kill the plant, but does make it malnourished because radiation damages chloroplasts, the photosynthetic cells in leaves that transmute sunshine into sugar, leading to less intense photosynthesis.

Very high radiation doses kill chloroplasts altogether, which is what happened to the Red Forest.

"The trees that stood closest to the reactor died first. In the first days, their exposures were as high as 8,000 roentgens an hour," said Svitlana as we strolled amid the stunted pines, passing a tall and skinny sapling that was utterly bare of branches but for the top, where three shoots had sprouted.

"The exposure decreased after the short-lived radionuclides decayed and those that remained washed off," Svitlana continued. "But a pine dies at an absorbed dose of 7 to 11 grays and the dose they accumulated over time was that high in places, so the Red Forest gradually expanded."

Radiation terminology is complicated enough for the layperson without having to contend with two different systems. Grays are metric units. Although metric units are not in much favor in the United States, except among scientists, they are the only units used in Chernobyl research.

The sievert—named in honor of Rolf Sievert, the Swedish scientist who did much to standardize radiation doses—is the metric equivalent of the rem. One sievert equals 100 rem. The "gray" is the metric unit for measuring absorbed dose. One gray is the same as 100 rads. So, if the Red Forest absorbed 7 to 11 grays, that was equal to 700 to 1,100 rads, a fatal dose. But the dose may not have come just from the radioactive debris and dust. It is possible that the steam that caused the explosion condensed in the cooler air, drizzling radionuclides onto the trees.

The trees' buds died first, followed by the cambium and needles. The chlorophyll and its green color were also destroyed, giving the forest its color—and its name. After the Red Forest was buried in the summer of 1987, some plants such as bison grass—an herb used to flavor *zubrovka,* or "bison vodka"—survived the interment and grew back the following spring, but their seeds were sterile.

Pine trees continued to die over the next several years as trees farther away from the reactor gradually accumulated lethal doses.

Even before the Red Forest's burial, surface radiation levels in the 10-kilometer zone fell dramatically by the autumn of 1986. Completion of the Sarcophagus put a cap on the lethally radioactive core. Short-lived radionuclides had decayed, bringing hourly exposure levels down to single-digit milliroentgens in most places outside the nuclear station. Though wind and rain are not very good at cleansing plants of hot particles and other fallout, the autumn leaves fell to the ground as the growing season came to an end. Except for evergreens and the trunks and branches of trees and shrubs, much of the Chernobyl fallout wintered on the ground.

SPRING CESIUM

The most direct route from Kiev to Chernobyl is a one-lane road that runs alongside the artificial reservoir known as the Kiev Sea and passes though village after village of small houses, hidden behind tall barriers and surrounded by fruit trees whose pastel blooms clothed the spring landscape in delicate hues of pink, ivory, and lavender. On each village's outskirts, the giant fields of industrialized agriculture were bordered by small rectangular plots, one right next to the other, where the farm workers kept their private gardens. The variegated patterns of each individual vegetable patch blanketed the land like a textured quilt.

Recently returned from their wintering grounds in Africa, white storks glided over the fields, their long red legs trailing like ribbons. Just a few miles outside the zone's southern checkpoint in the village of Dytiatky, I spotted a colony of 20 storks—the most I had ever seen in a single place—dining in a plowed field. Storks are practically sacred in Ukraine, and their large stick nests bestow favor, fortune, and fertility on any household lucky enough to have one perched on its roof, telephone pole, or chimney. But woe awaits the family whose storks abandon it.

Perhaps that is why white storks have become a symbol for Chernobyl, used to decorate posters and pamphlets about the disaster. In contrast to their shy and rare black stork cousins, white storks like living amid people, in open farmlands near swampy riversides, marshes, and floodplains stocked with stork snacks such as frogs and snakes—and just like the lands around Chernobyl on the eve of the disaster. But these are now no-man's-lands.

I thought about storks in the abandoned village of Leliv, which overlooks the Chernobyl station's cooling pond just inside the 10-kilometer zone. It wasn't an easy place to reach. Though I was able to drive part of the way on crumbled asphalt through a dense thicket of branches that slapped against the car windows, my companions and I had to get out and walk when we came upon a young aspen that had sprouted in the middle of the road.

Leliv is a moderately contaminated pink patch on the radiation maps and my dosimeter displayed numbers well below its two-milliroentgens limit: 45, 67, and 23 microroentgens an hour. But I couldn't keep my eyes on it for long because I had to watch my feet. A thick carpet of grasses sprouting with jolly dandelions and plowed into uneven clumps by boars made the going difficult. We couldn't stop because as soon as we did, swarms of nasty gnat-like bugs enveloped our faces. They didn't bite, but they were annoying as hell.

After about 10 minutes, we came upon the rotting remains of log cabins. Some of the roofs were of corrugated metal. Others had been thatch, but all that was left were some wisps of dried grass stuck to exposed beams. Part of a stork's nest rested on a cottage gable, though a large chunk had fallen off onto a jumble of rusted machinery, overturned carts, and other detritus left behind in the evacuation. When the people left and friendly farmlands steadily gave way to feral fields and wormwood forests, the storks eventually left, too, and their nests largely rotted, the debris blown away by wind.

Storks can occasionally be seen in the villages where people still live, though not many, and there are none at all in completely abandoned villages like Leliv. After more than a dozen trips to Chernobyl, totaling about a month's time altogether, the only stork I ever saw in the zone was painted on a sign that the Chernobyl forestry service planted on a roadside.

A small, dark gray bird with a rust-colored tail and a white wing patch was perched on the cabin's window frame. It was a black redstart. The birds colonized the abandoned houses after the evacuation.

"This all used to be cultivated," said Svitlana, waving her arm at the wild fields, thick with bushes, shrubs, and small trees that were consuming the village. "But nature started taking over as soon as people left. The environment is returning to the way it was in the sixteenth century."

The forests that covered 80 percent of Polissia in the twelfth century and gave the *drevlyanny* their livelihood and name were gradually logged over the centuries. By the sixteenth century, only half of Polissia was woodlands. In the nineteenth century, swaths of the region were denuded for lumber and some swamps were drained. The Soviets drained even more swamps and planted trees on about a third of the land after World War II, but they were pine plantations, devoid of diversity. Most of the remaining land was cultivated, although it was hard to believe that the blooming springtime wilderness surrounding us had ever been under the plow. The 1986 fields, sown just before the evacuation, grew without anyone to tend or harvest them, but there wasn't a trace of cultivated crops that I could see in Leliv.

When we made our way through a stand of young silver birches, their yellowish catkins just beginning to open, I heard what sounded like a kitten quietly meowing. It was a jay, singing on a birch. A big tawny bird with a wing patch of iridescent blue, the European jay's call is crow-like squawk, though its song is an unmelodic and odd mixture of clucks and meows. It was the first time I had heard it, but when I focused my binoculars on its source, the bird had already flown away.

"Once cultivation stopped in the 30-kilometer zone, succession started almost immediately with wild grasses, herbaceous plants, and shrubs. After 15 years, the meadow stage is completing and tree seedlings are becoming more frequent," Svitlana continued.

Succession refers to the changes in a plant community that occur over time, creating and filling gaps created by natural (floods, windstorms, volcanic eruptions) and unnatural (cultivation, urban development, swamp drainage) disturbances. A cleared woodlot is quickly colonized by the trees that remain in the vicinity. A pasture can eventually give way to a forest. Wetlands are re-created when neglected drainage ditches get clogged with silt. Given that the 30-kilometer zone may be uninhabited for about 300 years—the 10 half-lives that it will take for cesium and strontium to decay to relatively safe levels—it will have plenty of time to revert to its swampy woodland origins.

Actually, Chernobyl released several cesium and strontium isotopes with different rates of decay. Cesium-134 has a half-life of about 2 years, while that of cesium-137 is 30 years. Strontium-89's half-life of 50 days is much, much shorter than strontium-90's 29 years. At the turn of the millennium, only cesium-137 and strontium-90—abbreviated as Cs-137 and Sr-90—were of any concern. So when I refer to "cesium" or "strontium" without any numbers, I have in mind cesium-137 and strontium-90.

Because of the plutonium-239 that fell closest to the reactor, the 10-kilometer zone will be uninhabitable for all imaginable time. The half-life of plutonium-239 is 24,110 years—about the same amount of time that has passed since modern humans first occupied Polissia during the last Ice Age. Thirty plutonium half-lives amount to a staggering 723,300 years.

"And how about radiation?" I asked Svitlana.

"Nature doesn't notice radiation—not at these levels. At least not in a way that's obvious to the eye," she replied, sweeping her arm out to her side. "But all these plants are radioactive."

In contrast to the radioactive releases in the civilian and military nuclear industries, Chernobyl was unique in the annals of radioecology not only in the magnitude of the release but also in the variety of physical and chemical forms that the release took. For example, during atmospheric testing, up to 90 percent of the radioactive cesium and strontium formed when the radionuclides that vaporized in the intense heat of the nuclear explosions condensed in cooler air. Such so-called condensates are water soluble and exchangeable, which are precisely the chemical forms that can be taken up into living things.

Chernobyl also released condensates, but they were light and were carried by the wind as far away as Ireland. Most of the contaminants that got dumped on the 30-kilometer zone were not condensates but radionuclides embedded in "hot particles"—insoluble bits of fuel, fission products, and radioactive graphite that were expelled from the reactor during the explosion and ranged in size from motes to chunks. In England, nearly all of the cesium was condensate. In the zone, nearly all of it was in hot particles. That's why lessons learned about the health and environmental effects of nuclear weapons testing are not always relevant to Chernobyl.

Moreover, the other radionuclides spewed on to the zone were almost all embedded in hot particles—zirconium-95 and molybdenum-99, as well as several isotopes each of europium, neptunium, curium, americium, and plutonium. But because such elements don't play a role in biology, not much of the precious government (domestic and foreign) funding is allocated to researching their behavior in the wild. Nevertheless, a Ukrainian study in 2000 estimated that the zone contains 378 curies of europium-154, 194 curies of plutonium-238, 405 curies of plutonium-239 and 240, and 486 curies of americium-241. These are very large amounts. But it is worth comparing them to the figures for radiostrontium—20,790 curies and radiocesium—70,000!

During the early postdisaster period, all of that radioactive stuff was on the surface of things. Now after 18 years, 95 percent of all the radionuclides have sunk into the ground and from there some have made their way into plants. So, in contrast to the early postdisaster days, when nearly all plants' exposure was from external sources of radiation, more than half of their exposure now is from radionuclides that they have incorporated internally.

Unlike external exposure, the impact of internal radiation depends on an enormous number of factors, including the types of radionuclides that are internalized, which determines whether the radiation emitted is alpha, beta, or gamma.

"Can we measure it?" asked Rimma, taking my dosimeter.

Svitlana flashed a white smile and shook her head. The botanist had bright blue eyes and, in contrast to the usual camouflage, wore a blue jacket to match them. My eyes are green with a touch of hazel and went well with the camouflage outfit I purchased at an army surplus shop in Kiev. In part, I did it to save the 15 dollars a day it cost to rent clothes from Chernobylinterinform. Mostly, though, I wanted clothes that fit me.

Actually, I could have worn any kind of clothes in the zone so long as I didn't mind leaving them behind in the unlikely event that they got contaminated dust on them. But I quite liked my Chernobyl commando outfit and fancied that I resembled Linda Hamilton's character in the movie *Terminator 2: Judgment Day* whose nuclear theme also seemed apt.

It is worth mentioning here the "lead suits" that many people believe provide the best protection against penetrating gamma radiation (alpha and beta particles are more easily blocked). I used to be one of them and once asked Rimma why there aren't lead suits at Chernobyl. She reminded me that lead is very dense and heavy. Clothing could only have a very thin layer of the metal, otherwise it would be impossible to move. A layer that thin would block only about half of the

gamma rays and would nevertheless slow down your movements so
much that you would increase your exposure just because you wouldn't
be able to dart in and out of the radioactive area quickly.

Rimma pointed the dosimeter near a wild plum. Adorned in its
spring garb of white flowers, the tree's thorns identified it as a wild—
rather than abandoned—plum. Cultivated plum trees are usually un-
armed.

"There's no change in the reading," she said with a puzzled frown.

"The dosimeter measures background radiation in the air,"
Svitlana explained. "It tells you nothing about internal radiation. For
cesium-137, you can determine internal contamination with gamma
spectrometry. But for strontium, americium, and plutonium, you have
to burn the sample and do a chemical analysis."

Because it's impossible to burn and analyze everything living in
the zone, no one knows the exact amount of radionuclides that have
gotten into the food chain, and estimates can vary considerably. But
even these are dynamic figures. Life is not constant. Radionuclides are
constantly flowing from one organism to another up the food chain
and from one ecosystem to another.

Svitlana pulled a needle off a normal-looking pine sapling and ab-
sently rolled it between her fingers.

"We've studied radiation's effects on pine trees the most because
they're so common here," she said. "Their internal radioactivity fluctu-
ates seasonally. It's highest now, in the spring, because the trees' juices
are activating."

Indeed, because the disaster happened in the spring, the plants'
fast growth promoted higher radionuclide uptake than if it had happened in the fall or winter.

Different trees favor different radionuclides. The Ukrainian zone's
populous pines contain more cesium than strontium. Birches have
more strontium than cesium, as do spruce, oak, birch, and aspen.

But just because a radionuclide is in the ground doesn't mean that
it is bioavailable, or capable of being assimilated by living things. First
of all, the nuclide must resemble something that organisms need. Plutonium, for example, imitates no atom of any use in biology and little
of it gets into plants through their roots. Cesium, in contrast, is chemically similar to potassium, and strontium mimics calcium. Both are
essential plant nutrients.

Potassium plays a role in synthesizing proteins, transporting sugar molecules across cell membranes, and regulating how much water is lost on leaves. Calcium, among other things, affects the movement of chromosomes. So, when cesium-137 and strontium-90 chemically confuse plants into using them instead of the needed potassium and calcium, the radionuclides are shunted to those cells where calcium or potassium are needed at that time, exposing whatever biological molecules are nearby to highly localized beta and gamma radiation.

Light and fast, the beta particles emitted by cesium-137 and strontium-90 travel farther in tissue than alpha particles, traversing nearby cells and ionizing atoms along their path by knocking off electrons and breaking chemical bonds. Depending on where these atoms are and what they are doing in the cell, the results can range from harmless mischief, to cancerous mutation, to fatal damage.

Because they are so penetrating, gamma rays—emitted by cesium-137's decay product barium-137—can do more distant damage than alpha and beta particles. But because living tissue—like most matter—is largely empty space at the atomic level, and gamma rays are subatomic in size, a lot of gamma radiation zips through without actually interacting with anything. What does, transfers its energy to subatomic particles such as electrons, which get very excited, breaking off whatever atom they were attached to and zapping around much like beta particles.

By 2003, nearly all of the external radiation bombarding us during our stroll comprised gamma rays, from barium-137 in the soil surface. But the gamma radiation decreases slightly each year as the cesium (and barium) decay.

Plants can't take up all forms of cesium-137 and strontium-90. They can only absorb chemical nutrients as ions, which are positively or negatively charged "free" atoms, or as simple complexes, such as sodium chloride or table salt. Since living things are mostly water, the chemicals they need for nutrition must also be water soluble, like salt, which dissolves into sodium and chloride ions. Insoluble things don't dissolve in water and can't be absorbed.

This is why trees that grew on contaminated patches of Waterford, Ireland, in 1994 absorbed 13 times as much of the cesium deposited there as trees growing in Kopachi, in the 10-kilometer zone. Ireland was contaminated by condensed, water-soluble cesium carried large distances from the reactor. Kopachi's cesium, in contrast, was mostly in fuel particles. So, even though there was much more radiocesium in Kopachi soil than in Waterford, a smaller proportion got into the plants.

Although 95 percent of all radionuclides are believed to be in the upper layers of soil, bioavailable isotopes of cesium and strontium are more likely to be found in the zone's living things. But strontium is more biologically active. While about one-fifth of the strontium is believed to be in zone vegetation, only about three percent of the cesium is thought to be.

Indeed, one of Chernobyl's many surprising lessons is that just knowing how much radioactive cesium is in a given patch of land will tell you nothing about how much of it is in the vegetation growing there. This depends on whether the cesium is condensed or embedded in a hot particle and whether is it soluble or not. Soil quality also plays a critical role.

Radionuclides don't sit still. They migrate, or move, from one place to another. Charts of cesium's flow through an ecosystem are extraordinarily complex filigrees of knitted ties and connections. The nuclides often migrate with percolating water—which is how they get into plants—and the type of soil is largely what determines how much. Cesium gets into plants the most in peat lands.

Peat is made of dead sphagnum moss that has not fully decomposed because the environments in which the mosses grow are low in nutrients, oxygen, and the bacteria responsible for decay. Waterlogged peat lands, called mires, form either bogs or fens. A bog is fed almost exclusively by rainfall and other forms of atmospheric precipitation. It is very low in nutrients and supports few species of plants, such as sphagnum mosses as well as carnivorous plants that must get their nutrients from insects. Things that fall into bogs barely decompose, which is why there is an entire subdiscipline of archaeology devoted to bogs. A fen, in contrast, is fed by groundwater and river flooding. Because it is richer in nutrients, sedges, reeds, and many other plants grow in its shallow waters.

Cesium is very mobile in mires because the partly decomposed dead moss that is peat's primary ingredient retains water and the cesium keeps floating around. Floating cesium is generally soluble cesium—precisely the kind that gets taken up by plant roots. Also, peat soils are acidic, which also makes cesium more water soluble. This explains why the Kiev region of central Ukraine is more contaminated with cesium, but plants growing on the drained, acidic peat lands found in parts of the Polissia region incorporate more of the radionuclides internally.

Indeed, it became clear after atmospheric nuclear testing in the mid-1960s that Polissia was absolutely anomalous when it came to radionuclide mobility up the food chain and into people. Given the same levels of soil contamination after global testing, a person living in Polissia absorbed 10 times as much radioactivity as a person living in Moscow or Minsk—all because of the type of soil there.

Cesium migrates the least in soils with a lot of clay because the minuscule clay particles are very sticky and the cesium gets glued to them, becoming increasingly biounavailable with time.

Or so, at least, it was thought. It may be that the cesium is being fixed by bacteria rather than clay. Some species of bacteria have such a huge appetite for cesium—any kind, radioactive or not—that their ability to incorporate the stuff is comparable to laboratory materials specially developed to separate cesium from other elements. Moreover, the cesium-eating strains can be found anywhere their favorite food is. Samples of the microbes taken from close to the ruined No. 4 reactor are almost as radioactive as nuclear fuel. The bacteria hold promise in radiological cleanup, although such applications require a good deal of work. For one thing, when they are on surfaces, the bacteria are basically indistinguishable from dust, so they can spread radioactivity as well as fix it.

Thus far, however, no one has identified strontium-eating bacteria. In fact, strontium doesn't fix onto much of anything in soil. In contrast to cesium-137, the amount of water-soluble strontium-90 in the root level of a given patch of soil is a pretty good measure of how much strontium will be in the plants growing there. Unlike cesium, whose absorption by plants was stabilizing after 15 years, the amounts of strontium continued to grow.

SUMMER STRONTIUM

July 6 is a magical date in Ukrainian folklore. Popularly called the Eve of Ivan Kupala—which is a merging of St. John's (or Ivan's) Day in the Orthodox Church calendar with Kupala, the name of a pagan Slavic fertility godlet—it is a night when the sun was said to bathe by dipping into waters on the horizon, imbuing streams and lakes with special charms. A wreath cast into their currents will presage a maiden's marital prospects. Herbs gathered before sunset are especially powerful, while at night all plants are able to walk and talk. It is also a dangerous night, when sundry demons and sprites prowl for human victims. Considered especially dangerous were the *rusalkas,* water nymphs, whose fear of wormwood prompted the superstitious to hang sprigs of the weed on their cottages and outbuildings. Even Soviet disapproval couldn't entirely eliminate the folk customs, including the charming Polissian spring ritual of "banishing the *rusalkas,*" in which a girl or female effigy was ritually buried in a field by a river, symbolizing an annual cleansing of the evil spirits. But *rusalkas* revived their powers in the magical waters of Ivan Kupala.

Actually, I had never understood why *rusalkas* were believed to pose a problem in a region where wormwood has been common since the Ice Age and the largest town—Chornobyl—was named after the plant (for folklore purposes, *Artemisia vulgaris* and *A. absinthium* were both effective *rusalkas* repellants). But in a feral meadow of parched yellow grasses, blooming wildflowers, and the occasional *Artemisia* specimen in the southern quadrant of the 10-kilometer zone, I mused that perhaps *rusalkas* sought revenge for the gradual destruction of their woodland habitat. The field I was in had been forested for millennia, but it was leveled for lumber in the nineteenth century and in Soviet times became a potato farm. Before 1986 it would have been cleared of any vegetation that was not potato—including wormwood—and *Artemisia*'s absence made the land vulnerable to the dangerous nymphs. Now that nuclear wormwood from Chernobyl has put an end to cultivation, the landscape is returning to its natural state and the repellant plants have been able to return, once again making the zone a *rusalkas*-free habitat.

Such whimsical thoughts occurred to me as I followed Rimma and Svitlana the botanist on a short hike in Cherevach, an abandoned village on a floodplain of the Uzh River. The dirt path that led from the main road into the village wasn't much of a road. In fact, it wasn't a road at all. After about 20 feet, it ended abruptly in a sea of deep grasses and shrubs.

Ivan Kupala was a warm and dry day with a light wind, but the air was sweet from the previous night's rain. Parched conditions were precisely right for stirring up radioactive dust. Radiation readings are always slightly higher on dry, windy days.

Cherevach was on a dark-beige patch on the cesium maps, meaning it was only moderately contaminated with that radionuclide, although its strontium-90 concentrations were among the highest in the zone.

Ninety percent of the strontium that fell in the Ukrainian zone was embedded in the insoluble hot particles that have sunk into the top layers of soil over 18 years. There is less field research in Belarus than in Ukraine, although only about half the strontium there is thought to be in the fuel particles. Because they are insoluble, hot particles sink by mechanical mixing and migrate much more slowly than the condensed cesium and strontium that flow with percolating water and have sunk more deeply.

Strontium in a hot particle becomes bioavailable only when the particle dissolves. So the particles are like a line that prevents radionuclides from getting into the food chain. The soil largely determines how quickly hot particles dissolve—the more acidic the soil, the faster the disintegration. Given the variety of zone soils—sandy, peaty, podzolic—disintegration has been very uneven, although it was relatively rapid in the acidic podzol beneath our feet. Podzolic soils are formed by forests that grow on the deposits left by melting glaciers. Because it is acidic and has very little organic matter, podzolic soil is not fertile.

At first it was believed that all of the particles were bits of uranium oxide fuel, most of which would dissolve after 15 years and release the maximum amounts of strontium in 2006. By then a good part of the nuclides will have decayed, so the total amount of it released from the particles will be only about a fifth of the amount of "free" strontium in 2001.

But it may be that things are more complicated. Scientists who scanned Red Forest fuel particles with an electron microscope found that there were actually three types of particles. The first group comprised bits of uranium oxide that broke out from the reactor core in the explosion, the second was made up of bits of uranium oxide that burned in the graphite fire before being thrown from the reactor, and the third group was made of bits of uranium oxide that had not burned in the graphite fire but instead had melted together with some of the zirconium from the fuel rods.

Although the burned bits of fuel were the least stable—dissolving easily—and the unburned uranium oxide was quite stable, the scientists were surprised to find that the zirconium particles were superstable. Since nearly all of the zone's strontium is in various kinds of fuel particles, the fact that not all of the particles dissolve very easily means that the amount of strontium that will eventually become bioavailable may be less than previously thought.

For more than a decade after the disaster, scientists had only a vague idea about how much strontium there was. The isotope emits no gamma rays, and its relatively low-energy beta particle doesn't help identify it. First of all, beta radiation is stopped by solid materials, and if the beta's source is in the soil, there are plenty of solids there to stop much of it from reaching the surface. Even if a beta particle got through, however, you wouldn't know if it was from strontium-90 or from a natural beta emitter such as carbon-14 or potassium-40. Unlike alpha particles, which have the same energy for any given isotope, beta particles have a range of energies up to a certain maximum. The only way of knowing how much strontium-90 is in something is to do expensive, laborious, and time-consuming reactions to separate the isotope from other elements and then, finally, measure its beta decay.

A Belarusian scientist once told me that doing strontium chemistry was "women's work" compared to cesium. Cesium chemistry merely involved putting a sample into a machine, which measured the gamma radiation of its decay product barium-137 and popped out an answer. In contrast, just separating strontium from whatever matrix it is in requires a great deal of mixing, heating, centrifuging, and adding ingredients—all done in a beaker on a stove, just like a "hausfrau" laboring over dinner.

Mapping the zone's strontium required doing that many times with many samples taken from many places. When Ukraine finally did this in 1997, scientists concluded that most of the Chernobyl strontium was condensed and had floated beyond the zone's borders. As for the hot particles that fell within its borders, they were very unevenly distributed. Nearly 80 percent are concentrated on one-tenth of the territory.

The profuse grasses were incredibly thick, and our walking stirred up a variety of insects that flew up my nose and stuck to my sunglasses. Swarms of mosquitoes bit imperviously through the OFF I had smeared on my body and sprayed on my clothes before setting out. Polissia's summer landscapes were picturesque, but the mosquitoes were punishing. I slapped them, wondering if they were radioactive

and, if so, was every dead mosquito on my skin leaving a patch of contamination. But Svitlana was a botanist and confessed to knowing
nothing about mosquitoes.

Rimma was scouting up ahead with my dosimeter and calling
out radiation levels. We were on a berry hunt, but it seemed none of
us were very good at it. A patch of huckleberries we found hadn't rip-
ened yet.

"We need to find an old garden or something," said Svitlana.
Though berries grew wild in the forests, most Polissians also planted
them in their gardens for easy access. "Raspberries and blueberries
should already be ripe."

The trouble was we couldn't find any. We weren't looking for berries to eat them. In fact, the first sign you see upon entering the 30-
kilometer zone is a large billboard explaining that open fires, hunting
and fishing, and wild mushroom and berry picking are forbidden.
I wanted to find some berries and then take them to a lab for strontium testing but making that a part of this story wasn't looking very
promising.

I asked Svitlana why berries are such magnets for radionuclides.

"Berries are usually shrubs or herbaceous plants," she explained.
"They have relatively shallow roots that penetrate into the layers of soil
with the highest concentrations of radionuclides. The root hairs that
absorb inorganic ions like calcium and potassium are especially concentrated in those levels. Plus, berries are reproductive tissues and all
reproductive tissues concentrate nutrients—as well as the radionuclides that imitate them."

A similar principle applies to grasses, which also have shallow root
systems. But people don't eat grass. They do eat berries. (They also eat
the game animals that eat grass and berries.)

I was hoping to find currants, which are especially radioactive. But
instead of currants, we found some dark green sorrel growing amid
the dry grasses. Like all leafy vegetables, sorrel was a good source of
calcium—and, in zone sorrel, strontium. But although strontium
chemically mimics calcium and plants metabolize them in similar ways,
cell membranes distinguish between them. Given access to both calcium and strontium, living things will preferably absorb the real nutrient. They will also select potassium over cesium. That is why fertilizing
cultivated fields with potassium and calcium decreases the plants' up-

take of radionuclides. But no one had fertilized the zone for nearly two decades.

Unlike potassium—and its imitator radiocesium—which both peak in the spring and then decrease, calcium and radiostrontium are stored by plants, which start accumulating the mineral in summer and continue to do so throughout the growing season.

"That's why you find more strontium in older plant tissues at the end of the growing season, in the fall," said Svitlana. "In pine for example, young needles contain more cesium than strontium, while the older ones have more strontium than cesium."

Another reason for this is that calcium (and strontium) are not very mobile in plants. Unlike potassium (or cesium), which can get transferred from place to place as needed by the plant's circulatory system, calcium (or strontium) stays put.

Rimma peered into the distance through a cloud of mosquitoes that hung in the air and announced that she saw what looked like an abandoned yard. She went ahead of us to reconnoiter and shouted out radiation readings that were double and triple natural background.

"Even if there are some berries left, it's unlikely we'll be able to see them in this deep grass," Svitlana said, brushing a black bug off her check.

I called out to Rimma who seemed to have disappeared into the thicket. But she emerged quickly, explaining that she had gone to take a closer look at what looked like raspberries only to find nettles instead.

It was a fool's errand to look for berries while tramping through a carpet of grass that was so thick it was like walking through deep snow. Svitlana pointed to the thatch—the deep, tightly tangled layer of living and dead stems, leaves, and roots that had accumulated between the layer of actively growing grass and the soil. She dug at it with her heel to expose some of the more rotten layers beneath the surface but didn't come even close to exposing the ground beneath. The thatch must have been about a foot thick.

"Radionuclides are less fixed in natural meadows and pastures than in cultivated fields because they accumulate in the thatch and remain bioavailable," she explained.

As a variety of invertebrates, fungi, and microorganisms decompose the thatch, the radionuclides locked inside the plants return to the soil. In plowed fields, thatch never has a chance to build up in the first place.

Understanding many of the "hows" and "whys" of radioecology will take many decades. One problem is money. Neither Ukraine nor Belarus has much of it for research at a time of budgetary shortfalls for even basic safety works in the zone, while international funding has tapered off significantly since the 1990s. Chernobyl has yet to generate more than sporadic science despite high hopes in the immediate aftermath of the disaster, which some scientists referred to as the world's largest field experiment in radiology.

Indeed, when Chernobyl exploded in 1986, American scientists knew exactly as much as they had known in 1976 about radiology because in the early 1970s the U.S. government cut research funding. Studies that started in the 1950s and 1960s and were designed to be long term were cut. But even those studies were usually laboratory experiments where a gamma radiation source was stuck in some area and scientists studied its effects on the surroundings. There were very few studies of the effects of fallout, and little was known about the effects of beta radiation.

Actually, it is more accurate to say that whatever was known was known to only a few. The Soviets had done radioecology studies after nuclear spills in their nuclear weapons industry. But these were super-secret and unpublished—a good deal remains classified. When those scientists came to the Chernobyl zone immediately after the disaster, they already knew much that was new to civilian scientists.

Chernobyl provided a unique opportunity to openly study all that and more, but it won't last forever. With biologically active radionuclides such as strontium and cesium constantly decaying, the window for studying high levels of contamination in the environment will eventually close. At the same time, Chernobyl's long-lived radionuclides such as plutonium and americium will continue to affect living things for a very, very long time to come.

The very fact that vegetation is thriving in the zone means that plants can adapt to living in a radioactive environment. But the plants are under evolutionary stress and intense natural selection. The result could be a selection for genes that are highly adaptive to stress, that could eventually be used in producing transgenic plants. But more time and study (as well as money) will be needed.

The zone's very uniqueness also poses a problem. If its plant life is considered part of a giant field experiment to observe the impact of radiation on natural habitats, then the zone plants should be compared with controls. But finding exactly the same plants growing in exactly the same conditions—except without radiation—is more easily said than done for the simple reason that every part of the Earth's surface is unique. There cannot be any other place like it on the planet if only because no other place can have the same geographical coordinates—to say nothing of the properties of soil (such as the anomalous ability of Polissia soils to transfer radionuclides into the food chain), air, water, sunlight, and myriad other variables that make up a particular habitat.

AUTUMN FOREST

The October day was bright and crisp as an apple, illuminating the scarlet, umber, and gold foliage that still clung to the trees after heavy rains a week earlier. The carpet of forest litter had dried on the surface but was still damp beneath, crunching and squishing beneath my feet as I hiked through a forest just three miles from the reactor complex.

I was tagging along with two dozen Chernobyl explorers—a multigenerational and enthusiastic crew of ethnographers, radiologists, chemists, and biologists sponsored by the Ukrainian Ministry of Emergencies. Their job was to collect, codify, and preserve the cultural artifacts left in the zone after its evacuation.

The only odd note was the armed policeman in a flak jacket.

"He's for protection," said Yaroslav Taras, an architect with salt-and-pepper hair who had appointed himself my escort and explainer. "There are outlaws who live in the abandoned villages, poachers, looters. There are hundreds of thousands of tons of abandoned radioactive machinery that people steal to sell as scrap metal."

With most of the 340 police and firefighters concentrated on the Ukrainian zone's perimeter, at the nuclear plant, and in the town of Chornobyl, vast regions are essentially lawless.

I thought of the barrows of Burakivka and all the contaminated waste that was just littered around the zone, waiting to be buried and pretty much free for the taking. The only problem was getting it past the checkpoints. In a poor and corrupt country like Ukraine, checkpoints are often merely a negotiating tool—although the Chernobyl checkpoints are tougher than most.

We were on the outskirts of an abandoned village called Novoshepelychi, a deep mauve patch on the radiation maps, and in the depths of an old forest whose border had been ringed with young pines. Trees, it turns out, are the best and cheapest way of fixing radionuclides because they stay put and decay away in the decades and centuries of a tree's lifetime.

Radioactive atoms generally target physiologically active tissues such as the cells responsible for photosynthesis in leaves and needles. A good portion of radionuclides are also in the bark. Since bark is not metabolically active, it may be because the contamination is external, from radionuclides kicked up into dust by the wind. With each annual layer of bark, older layers become trunk wood but, oddly enough, the nuclides are very mobile in wood and don't stay in it very long. The 1986 tree rings, which were bark in 1986 and coated with fallout, *don't* display the highest contamination levels. These are invariably in the cambium—the trees' circulation organ.

Once radionuclides are fixed in trees, the greatest danger in the wormwood forests becomes fire, like the 1992 blaze that consumed 30,000 acres (12,000 hectares) of forest and caused pockets of panic in Kiev. Burning breaks the chemical bonds locking radionuclides into biological molecules and resuspends them into the air much like the original Chernobyl fallout, although they don't drift very far. The main radiological danger is to firefighters, who risk inhaling plutonium.

For years, members of the expedition had been searching for the Novoshepelychi graveyard, but it was hidden so deeply in the forest that they hadn't been able to find it. But this time they had a guide, a stocky bus driver who grew up in the village.

The policeman in front with the bus driver, we walked swiftly up the path until we came upon the gravestones and metal crosses that seemed to grow out of the forest. The brittle whisper of falling leaves surrounded me as I tramped through the thick underbrush, my dosimeter hovering around a rather high hourly reading of 100 microroentgens.

When he saw me staring at the liquid crystal screen, Taras said: "Graveyards have higher radiation because the Polissians always put them on higher elevations, which caught more of the fallout. They knew that if the bodies were buried on lowlands, they would pollute the groundwater because it is so close to the surface in the swamps."

Zone cemeteries also have higher radiation levels because they are in forests and the trees are so big that they lock up a great deal of radioactivity. Of all zone workers, aside from those working with radioactive waste and those working close to the ruined reactor, forest rangers get among the highest occupational doses. In the early years, this was because radioactive grime coated the trees along their length, especially their crowns. In those days, the radiation dose actually increased the higher you went.

After a decade, one reason for the rangers' high exposure levels is banal—forest rangers are among the zone's most notorious poachers, and they also forage for mushrooms and berries—all of which contribute to their high average internal radioactivity levels. Another reason is that the forest floor is among the most radioactive parts of the zone environment.

Trees aren't the only forest vegetation to lock up radionuclides. Mosses are also culprits because, unlike vascular plants, they lack true roots and get most of their nutrients—and the radionuclides that mimic them—from the air. So, when wind kicks up radioactive dust or rain washes it off, mosses absorb the radionuclides easily. In 2001 a kilogram of some moss samples contained 90,000 becquerels.

The radioactive leaves that had been high in the air in spring and summer had fallen to the ground. While the expedition bushwhacked into the brush to videotape, photograph, and take notes on the gravestones, I made my way to an old oak surrounded by fallen leaves, many of them covered with galls. Most galls are formed by tiny dark wasps called cynipids, or gall wasps, which lay their eggs on different plants (each cynipid species prefers a different species of plant). The larvae release substances that cause the plant to grow extra tissue, some spherical and as large as golf balls, that serves the larvae as a nursery and pantry while they develop.

But some galls are symptoms of crown gall disease. This is caused by *Agrobacterium tumefaciens,* a bacterium that actually transfers a part of its DNA to the plant, which integrates it into its genome. In crown gall disease, this leads to the production of tumor-like growths that are the closest things to cancer in plants, although the growths usually don't harm mature plants. In plant breeding and genetic engineering, however, *A. tumefaciens* is widely used to insert useful genes such as those for herbicide resistance into plant genomes. Even under low radiation doses, the incidence of crown gall disease in the zone is much higher than outside, and the galls are much larger as well. This may be because radiation damages the plants' ability to recognize the crown gall bacterium and fight it.

Near the fence of an infant-sized grave marked with a tiny cross of welded metal, I kicked over a stand of mushrooms and dug through the forest litter with my heel, releasing a waft of fungal damp. It was a warm October, warm enough for nature's sundry sanitation workers to still be active. Centipedes quickly snaked away from the light and disappeared into the dark fragments of chewed-up litter from years past. Worms wriggled iridescently and millipedes rolled up into small pills protected by black armor.

The first layer of forest soil, known as the "O" or "organic" layer, is the recycling factory where decomposition releases the nutrients in dead forest products and makes them available for reuse. In Chernobyl forests, the O layer is the key to unlocking radionuclides.

The organic layer is made up of three easily distinguishable horizons. The top layer, called "Ol" for "organic litter," was the layer of freshly fallen leaves, twigs, branches, bark, flowers, fruits, and other botanical detritus that I was walking on. The "f" in the "Of" layer refers to the "fragments" of dark, partly decomposed plant debris beneath the fresh litter. The final layer was made of the amorphous and fully decomposed organic gook that is the prize at the bottom of backyard compost piles and the stuff that gives Ukraine's black earth its famous fertility. It is called "Oh," with the "h" standing for "humus."

The denizens of the decomposition from litter to humus are so sundry and numerous that science has yet to even count them all. For every creepy crawly I could see with my eyes, there were thousands more visible only under microscopes. In the first decade after the disaster, however, species diversity declined by more than half, probably because the forest litter was so highly radioactive. There were fewer ticks, millipedes, and other invertebrates in forest soil than outside the zone. But the populations have been rebounding in the second decade, perhaps because radiation levels have fallen or because the insects are developing resistance. Adult insects, as a rule, are highly radioresistant.

From teensy mites to even tinier bacteria and certain kinds of fungi, an astonishingly diverse and largely unknown world was at work. Indeed, so little is known about the microbial world that no one has ever documented the extinction of a single bacterium! Given such mysteries, it is impossible to know all of the pathways by which radionuclides migrate through the seasonal cycles of an ecosystem. But decades of study at other nuclear spills have mapped the basic mechanism, and

18 years of Chernobyl studies have revealed much more and suggested some promising avenues for research.

Along with bacteria, fungi are nature's principal decomposers, and they may play a surprising role in the movement of radionuclides through the environment. Occupying their own kingdom between animals and plants—though they are closer to animals than to green plants—the total number of fungi species may run as high as 1.5 million or more. But mycologists—or fungi experts—can't agree on this, or on the number that have formally been described, a figure that ranges from 74,000 to 300,000. And those that have been identified are hard to generalize. Some fungi, such as yeast, are single-celled. The rest are composed of filaments called hyphae that form a mass called a mycelium. The thread-like hyphae are so small that they can grow right through seemingly solid objects. Fungi don't ingest nutrients. They ooze enzymes onto wood, toenails, or some other substrate and then absorb the simple sugars and amino acids that are released.

Although we see only a part of them, fungi are such an integral part of the forest that if you removed all of the trees and soil and just left the fungi behind, you'd still be able to see the outlines of trees and soil.

As the mycelium webs out in the ground, no part of it is more than a few micrometers from the environment. When rain percolates through it, the fungal web acts as a filter, absorbing nutrients and water. In the zone, of course, the word "nutrients" should always raise red flags for the radionuclides that mimic them, and the mycelium's role as a nutrient sponge may explain one experiment in which the amounts of radioactive cesium in the Ol layer periodically *increased* even as the total amounts of litter *decreased*.

Forest soil and litter are, in general, very radioactive. There is more cesium in forest litter than in all other elements of wild flora. This is because the litter is made up of leaves and needles that bore the brunt of the 1986 fallout as well as fresh leaves—which also take up radionuclides—that fall each season. Since the concentrations of nutrients increase as organic litter decomposes to humus, so do concentrations of the radionuclides that mimic them. This is one reason the fragment and humus layers contain about a quarter more cesium than the fresh litter. Another reason is that nearly all of the zone's radionuclides are in the top two inches of soil, where the Of and Oh layers are located and also where fungal mycelia are found.

When fresh leaves fall, some fungal mycelia will grow their way into the new litter. If they have accumulated cesium from the Of and Oh layers, they will also import the radionuclide to Ol, increasing the concentration of cesium even as the total amounts of litter decrease. Since mycelia can also cover a very large area—some fungi are the largest living things on Earth—it may be that they can shunt cesium from far distances. At least this is one theory to explain the strange experimental results.

Nevertheless, fungi's role in the circulation of radioactive cesium is not well understood. One reason is that mycology is not exactly "sexy" science. The dank association of mushrooms with witchcraft, damp, and decay doesn't attract many young scientists. Radioactive mushrooms hold still less appeal.

The intimate relationship between fungal mycelia and the cesium-rich organic layers of the forest floor also explains why mushrooms contain more cesium-137 than any other forest vegetation. One of the highest concentrations ever was found in a highly poisonous naked brimcap (*Paxillus involutus*) growing on peaty soils 28 miles from the plant.

Mushrooms, in general, are highly contaminated. But much depends on the species and how deeply the main part of their mycelium penetrates the soil to layers where cesium is concentrated. For years after the disaster, mushrooms such as bay boletus (*Boletus badius*)—whose shallow mycelia don't penetrate past the humus layer—contained 10 times as much cesium as porcini mushrooms—whose mycelia penetrate past the humus into the mineral layers of soil. In 1997 a bay boletus mushroom found in the buried village of Yaniv, not far from the nuclear station, contained an astonishing 2 *million* becquerels of radiocesium per kilogram.

In Ukraine the maximum amount of cesium allowed in a kilogram of mushrooms is 500 becquerels.

Since 2000, however, shallow species like bay boletus have been decreasing their cesium uptake, while penetrating species such as porcini have been increasing it. In 2001 a bay boletus in Yaniv contained about 500,000 becquerels per kilogram and a porcini contained 160,000. Two years later, a kilo of Yaniv porcinis contained 500,000 becquerels of cesium, while the same amount of bay boletus contained only half as much.

Moreover, the amounts of strontium-90 in mushrooms have been increasing since 1996. In part, this has been a result of the disintegration of fuel particles, a process that is also aided by fungi. Strontium is much more dangerous to human health than cesium because it concentrates in bones, where its beta particles can damage bone marrow. It also has very low biological turnover. Whatever strontium you absorb stays in your body for years.

But even as porcinis' radioactivity levels are expected to continue increasing in the coming years, they don't win the dubious prize for cesium concentrations. This belongs to naked brimcaps, which contain 10 times as much cesium. But naked brimcaps are poisonous, and porcinis are highly prized by Polissian mushroom hunters.

The mushrooms growing around the graves in the cemetery were like the apples on a subterranean mycelium tree, bearing spores that are the fungal equivalent of seeds—except that the tree was huge, like a dense underground spider web that could extend for many square miles, so that the mushrooms I kicked over in Novoshepelychi might be siblings of mushrooms growing four miles to the west in Kopachi or four miles to the north in a patch of Belarusian forest.

It was hard to believe that Belarus was so close. Indeed, being in the Ukrainian part of the exclusion zone made it easy to forget that there is an entirely "foreign" part of the zone in a different country. And whenever I did remember, I simply thought that Belarus's zone was much like Ukraine's.

Until I went there.

3

Birding in Belarus

Was it true that bitter waters will reverse course and engulf the vault of the heavens?

Oksana Zabuzhko

The 90-mile stretch of the Belarusian-Ukrainian border that runs through the Chernobyl zone is one of the most radioactive parts of the planet, painted in the darkest shades of red on the contamination maps. A brown smudge of plutonium straddles the border just seven miles north of the nuclear station. But the border is not demarked in any way. There are no passport controls or customs inspections, and border troops refuse to patrol it without extra pay, which no one has ever offered them.

Two roads cross the border within the confines of the zone, which seem to make it easy to go from the Ukrainian portion to the Belarusian and back again. Actually, it isn't. Although Belarusians and Ukrainians don't need visas to visit each other's countries, anyone, regardless of nationality, requires permission to be in either country's exclusion zone. Visitors must also have an official escort. You can't just wander around by yourself.

For an American like me, who needed not only visas for both countries but also stamps in my passport to prove legal entry and exit, crossing the radioactive border was even more complicated. The most logical way to do it—just driving my usual route from Kiev to the town of Chornobyl and then crossing the border—was illegal since there was no one to stamp my passport at the actual border inside the zone.

If I was caught in Belarus without the proper stamp, I could be detained for three days and then deported. What I had to do instead was to exit Ukraine and enter Belarus at a border crossing *outside* the zone and then cross the border again, semilegally, *inside*.

The closest border point was on a bridge over the Dnieper River near the Belarusian village of Komarin, six miles away from Paryshiv, the Ukrainian zone's northeast checkpoint. The drive there took me only about half an hour out of my usual way, but the border took more than an hour because my car needed Belarusian insurance, whose acquisition required $20 and visiting sundry trailers parked on the river's mosquito-infested marshes. Uninsured foreign cars get confiscated in Belarus.

Just two miles after crossing the border, I was inside the 30-kilometer zone, only now I was in the Belarusian section. In fact, about one-third of that original zone was in Belarus, but it was very different compared to Ukraine. For one thing, you can enter the zone near Komarin without passing through any checkpoint. The road, at least, is unrestricted, and on the radiation maps, the area is virtually clean. Such clean spots are in the Ukrainian zone as well. In the rush to establish an evacuation zone in 1986, it was impossible to map radioactivity levels in any detail. So, a rough circle was drawn around its perimeter even though some lands inside were quite clean, while patches outside were dirty. But access to even clean parts of Ukraine's 30-kilometer zone was nevertheless restricted because the Sarcophagus, the nuclear plant, and hundreds of nuclear waste dumps inside posed continued security risks.

The barbed wire fence around the zone's perimeter was incomplete and didn't extend to the southernmost bit of Belarusian territory that seems to jut into northern Ukraine on maps. In fact, although it was within a 30-mile radius of the power plant, that part of Belarus was populated and not part of the exclusion zone. Instead, Belarus's exclusion zone extended for a radius of 50 kilometers to the north because that's approximately how far areas of high contamination—40 curies or more of cesium-137 for each square kilometer—extended from the nuclear station.

In still-Soviet 1988, 500 square miles of those lands were put aside in the Polissia State Radiological and Ecological Reserve (PSRER). More lands were added in 1993, expanding the PSRER to more than 830 square miles.

Ukraine, too, attached a ragged 200-square-mile tail of territory to the southwest part of the 30-kilometer zone, creating a single administrative unit called the Zone of Exclusion and the Zone of Unconditional (Mandatory) Resettlement. That unwieldy name is shortened to an even more unwieldy acronym that transliterates into English as ZEZU(M)R. The Cyrillic original—*3B3Б(O)B*—is used in so many documents and signage that it is worth a mention.

Together, the adjoining portions of Ukraine's ZEZU(M)R and Belarus's PSRER immediately around the reactor created a roughly oval-shaped no-man's-land totaling 1,838 square miles. This is almost twice the land area of Rhode Island and about half the size of Yellowstone National Park.

RADIOLOGICAL RESERVATIONS

Driving west across Belarus on the way to Chernobyl took just 15 minutes. There was only one road, and it ran along the border of the radiological reserve on the right. The reserve was not fenced. The roads and paths that led into it were marked by large rocks painted with trefoils and yellow signs hand-stenciled with red lettered warnings:

PSRER
Walking or Driving is
Forbidden!
Fine
up to 10
minimum
wages

Ten Belarusian minimum wages was about $80. This is a stiff fine for a poor country where the average annual wage in 2000 was $700. It was only $300 more in Ukraine.

Instead of heading in the direction of the reserve, I kept driving west towards the border because I wanted to drop off my car in Chornobyl—back in Ukraine—before going farther into Belarus.

It would have seemingly been more logical, legal, and easier for me to simply drive from the border to the administration of the radiological reserve in Khoiniki, a town of 19,000 about 40 miles north of the Chernobyl power plant. But I didn't have a Belarusian map with enough detail to get me through the rural, sparsely inhabited, and poorly marked corner of the country I needed to drive through.

Although I had cobbled my own map together from inexpensive Ukrainian topographical maps that showed much of the bordering regions of Belarus, a critical corner of 15 square kilometers was missing, and it was the very area that I presumed must contain the turns to Khoiniki.

I consider myself fairly intrepid. But I was not so adventurous (or foolish) as to drive around a radioactive zone without a map, especially without certain access to unleaded gas for my car. Though I was quite certain there was unleaded gas, I knew that I would have problems actually finding it in a remote rural region.

It was surely the center of the universe to anyone who lived there, but it was the middle of nowhere to me and I needed both a guide and a ride. Once in Belarus, I hoped that I'd be able to buy a detailed topographical map with the sections I was missing. In the meantime, I wanted my car in a safe and familiar place. (Yes, given enough time, even a radioactive town like Chornobyl can seem safe.)

Rimma Kyseltsia, who came along as my companion and fixer, had arranged with Petr Palytayev, the director of the reserve, to pick us up at the Paryshiv checkpoint. But he wouldn't leave his office until he knew we had arrived in Chornobyl. It was an hour's drive from Khoiniki to Paryshiv and since he didn't have permission to enter the Ukrainian zone, he'd have to wait in his car outside the checkpoint where, to put it mildly, there is not much to do.

And that just about exhausted all the information we had about what awaited us in Belarus. We didn't know where we'd be staying, what it would cost, or even what we'd be eating. Unlike Ukraine's Chernobylinterinform, the Belarusian reserve's administration didn't seem to have a set price list for visitors. Palytayev had been quite cagey about all the details.

"We'll figure all that out when you get here," he had said when Rimma asked about prices. That could mean it would be free of charge, with a nominal gratuity to Palytayev. Or it could mean a shakedown. I have experienced both after being told variations of "oh, let's not discuss tacky issues of money now" in this part of the world.

I had visited the Ukrainian zone so many times that I had a blasé been-there-done-that breeziness about it. I knew where the shops were, what I could buy there, and what I needed to bring. In Belarus, I knew nothing and so prepared for the worst.

I folded emergency hundred dollar bills into the demi-pad pockets of my bra, packed an insulated bag with enough food to last three days, and even lugged a five-liter bottle of water along for the trip. Since alcohol is often decent currency in such parts and I had no idea of whether I'd even be in the vicinity of a store to buy any, I also brought along vodka and brandy.

The fact that rain poured from gloomy skies for much of the drive didn't alleviate my reservations. Belarus was a daunting, neo-Soviet dictatorship. While Ukraine was not exactly an exemplar of democracy, it had declassified hundreds of previously secret KGB documents about Chernobyl in 2001. Belarus's Chernobyl files largely remained secret.

In any event, it was good that I brought the food—and the booze.

After driving about 15 minutes through Belarus, we passed the tiny village of Gden. Little of it was evident from the road except for some aged log cabins. But they were quite clearly occupied. Gden was about the same distance from the nuclear power plant as the town of Chornobyl. But it was successfully decontaminated and, therefore, never evacuated. In Soviet times, Gden was cited as a positive example in contrast to the Ukrainian republic, which didn't decontaminate any villages in its portion of the 30-kilometer zone and simply evacuated them.

While Kiev did decontaminate the towns of Pripyat and Chernobyl, Belarus decontaminated the larger town of Bragin, just outside the northeast border of the radiological reserve, without ever evacuating the people (though all the children were sent away to summer camps).

Within minutes, we came upon the international border, which was marked only by a cluster of signs on either side of a candy cane post. Belarus displayed a small red sign that read: Border—Ukraine. The Ukrainian sign announced that we were entering the exclusion zone and warned us against picking mushrooms or berries. It was a low-key display compared to the border crossing at Komarin, where the signs scream and huge national flags whip in the wind.

About a minute later we were at the Paryshiv checkpoint on the zone's northeast perimeter, just four miles from Chornobyl.

At that point, I was in Ukraine illegally because I had reentered the country without an entry stamp. But there were no border guards or immigration officials in the zone to notice or care. The only thing the guards at the Paryshiv checkpoint wanted to know was whether I had a

Chernobylinterinform program, which was my permission to be in the Ukrainian exclusion zone. They didn't even ask to see my passport.

Once we arrived in Chornobyl, Rimma called Palytayev to start heading our way. After a lunch of sandwiches and fruit from my insulated bag, I parked my car in a garage, and a driver for the Chernobyl Ecological Center gave us a ride back to Paryshiv where Palytayev and his driver were waiting in a gray Volga sedan.

A minute later we were back in Belarus—where I was again somewhat illegal since I had recrossed the border without a stamp. But since no Belarus immigration officials would possibly find out (until and unless they read this book), the entry stamp I got in Komarin a few hours earlier was good enough to prove legal entry.

Rotund and mustached, Palytayev was a former Communist Party functionary who worked in local Belarusian government before taking over as director of the Polissia State Radiological Ecological Reserve a year earlier. He said that the reserve employed 700 people—firemen, checkpoint guards, forest rangers, security guards—nearly all of whom lived in Khoiniki and were bused to the reserve daily.

Though it is officially a radiological reserve, its workers still call it the zone in casual conversation. I also use "zone" to refer to the no-man's-land surrounding the nuclear station as a unitary radioactive environment, though I use each country's proper name for its respective portion of evacuated territory when necessary to distinguish them administratively or politically.

Seven hundred workers in the Belarus reserve amounted to far fewer than the 2,500 people who worked daily in the comparable, if slightly larger, Ukrainian exclusion zone. (And this doesn't count the 4,000 who worked at the nuclear plant.) The reason for the difference is that Belarus's reserve is a radioactive wilderness and nothing more. It requires little in the way of maintenance except fire prevention and security patrols.

Nearly 300,000 people passed through the zone, working short tours of duty, at the height of the cleanup in 1986 and 1987. By 1989 the total number of cleanup workers, known as "liquidators," approached 900,000. But after that, when the main cleanup work was done, the numbers dropped radically. In 1991, when Ukraine took over its portion of the zone, about 11,000 people were working in the administration of the zone and the power plant. By the turn of the millennium, there were fewer than 8,000 and the number continued shrinking after the plant was shut down in 2000.

So, what do they all do? Aside from the nuclear power plant and Shelter Object, people work in managing radioactive wastes, water resources, and forestry. There are police and firemen, scientists, construction workers, and medics—plus the people who provide services for them, delivering and preparing clean produce from outside the zone, running Chornobyl's three bars and its grocery stores. Despite the layoffs, it still takes quite a few people to run a no-man's-land.

In the age of terrorism, security has also become a more driving concern. After all, Red Forest soil contains enough radioactive cesium, strontium, plutonium, and americium to make a "dirty bomb" in every sense of the word, to say nothing of the radioactive mess inside the Sarcophagus.

After driving in heavy rain for about half an hour, splashing through puddles past clusters of dark wooden cabins with bright blue or green window frames, we drove across the railroad tracks that shuttle between the nuclear station and Slavutich, the brand new town for the station's workers built 30 miles to the east to replace contaminated Pripyat.

"Ukraine pays for maintaining this track," said Palytayev. "After all, they're the only ones that use it."

Originally, the workers were housed in a hastily thrown up settlement called Zeleny Mys, just outside the zone's southeastern borders. The ill-fated site was littered with World War II mines that had to be exploded before construction could begin. When it was completed, little—including central heating—actually worked.

Slavutich was built to replace it in 1988. Like a company town once the company has left, the town of 26,000 has been struggling to find a new purpose since the nuclear plant closed. It is a low-slung place of young trees, dowdy apartment buildings, and an incongruously gigantic plaza that seems empty no matter how many people are in it.

In Soviet times, the fact that the tracks went from Slavutich in the Ukrainian republic, through a portion of the Belarusian SSR, and back into the Ukrainian republic to the power plant made little difference since Soviet republics' borders and budgets were mostly administrative fictions run by the Communist Party from Moscow. But since the borders became international and the budgets national (and full of holes), the Slavutich-Chernobyl shuttle had become a bigger headache, requiring various government agreements to smooth the Chernobyl workers twice-daily international crossings.

We were driving through the missing corner of my map when we passed a sign showing the turn north for Khoiniki. But the driver turned south instead, passing feral fields of emptied villages.

Like Ukraine, Belarus (and Russia) mandated resettlement from regions of high contamination where radiocesium levels are more than 15 curies per square kilometer or where inhabitants will get an "extra" dose of more than one rem (10 millisieverts) a year. The annual dose most people get from natural background radiation is from 0.1 to 0.6 rem, depending on where they live and lifestyle factors. So, an additional rem is from about 2 to 10 times that.

That natural background dose of about a tenth of a rem is also the current public dose limit recommended by the International Committee on Radiological Protection (ICRP), though it uses the metric sievert units instead of rem. In the past, the maximum lifetime dose was not supposed to exceed 350 rem, which would mean limiting annual exposure to 5 rem. So, the new threshold is much stricter.

Nevertheless, nearly all decisions on resettlement and compensation are based on the density of contamination in a particular region rather than the dose received by a person who lives there. That's because accurately measuring doses, especially low doses, is notoriously difficult.

For much of the 1990s, the highest priority had been resettling people from regions where cesium exceeded 40 curies and annual doses were more than 5 rem. Nearly all of these areas have been evacuated. Most, but not all, of these areas in Belarus were within the boundaries of the radiological reserve, which once contained 96 settlements and 22,000 people.

The second-highest priority has been resettling people in lands contaminated with 15 to 40 curies, but many people continue to live in these regions.

CALCULATING VICTIMHOOD

Palytayev gave me a pamphlet printed on the occasion of the reserve's fifteenth anniversary in 2003. Displaying the reserve's logo of a triangular trefoil sign against a backdrop of conical evergreens and oak leaves, the pamphlet was an informative primer for first-time visitors and one that the Ukrainian zone—which gets many more visitors than Belarus—would do well to emulate. But after more than 18 years, the

Ukrainian zone administration hadn't even put together a fact sheet, much less the colorful little booklet I held.

"The reserve contains 70 percent of the strontium-90 that fell on Belarus and 97 percent of the plutonium isotopes," I read aloud. "Exposure is as high as two milliroentgens an hour in places."

I had routinely exceeded two milliroentgens during my hikes in the Red Forest, where readings of 10 milliroentgens an hour of gamma radiation in the air are not unusual, and levels can go as high as one roentgen.

But comparing maximum exposure levels in the two zones led me to think about the various claims made on behalf of the three most affected countries—Belarus, Russia, and Ukraine—as to which country suffered most from the disaster.

It is often said that Belarus holds the unfortunate first place in Chernobyl victimhood. Indeed, Belarus officially maintains that 70 percent of the fallout from Chernobyl fell on its country. So do charities for Belarus's Chernobyl victims. For example, the web site of the New York-based Chernobyl Children's Project International (CCPI), which was associated with the 2004 Oscar-winning short documentary *Chernobyl Heart,* makes that claim and says that the reason is because the prevailing winds were directed north to northwest at the time of the explosion.

But Russian and Ukrainian scientists vigorously dispute Belarus's claims. They maintain that most of the radioactive release fell on the grounds of the power station and within the borders of the 10-kilometer zone in Ukraine. Although the 70 percent figure is repeated in all of Belarus's official documents and speeches, it is never footnoted or referenced in any way and even Belarusian scientists confess that they have no idea what its source is. A Belarusian radioecologist told me that he even made a special effort to find out where the number came from but got nowhere.

Instead, one international study that also included Belarusian scientists concluded that 33 percent of Chernobyl's cesium fell on Belarus, which is probably a good measure of the total amount of radionuclides that descended on the country.

One-third is nevertheless a large amount, but the explanation that so much fell on Belarus because of the prevailing winds is as unsatisfying as efforts to find the source of the 70 percent claims.

For one thing, the dirtiest territories were in the *eastern* part of Belarus, around the towns of Gomel—Belarus's third-largest city— and Mogilev, 180 miles to the northeast. But at the time of the explosion on April 26, 1986, ground-level winds were blowing south towards Kiev, while atmospheric winds a thousand meters high—where the graphite fire's smokestack effect lifted a good part of the radioactivity—went through *western* Belarus and arrived in northeastern Poland on April 27. From there the radioactive wind went on to Finland and Sweden, where it set off alarms at the Forsmark nuclear power plant north of Stockholm.

A light ground-level breeze shifted towards Gomel on the 27th, but the reactor's release of radionuclides had dropped significantly by then, as had the force of their expulsion from the core. So it seems unlikely that either was sufficient to blow the contamination to Gomel. Atmospheric winds shifted in the city's direction on the 28th, but by the time the release of radionuclides increased significantly on May 1, both ground-level and atmospheric winds were directed south. In short, wind direction does not explain the causes and extent of Belarus's contamination.

Rain was a likelier culprit. Soviet newspapers traced Belarus's contamination to heavy rains on April 28, which would have brought the radionuclides down when atmospheric winds were directed towards Gomel. Others maintain that the Soviet government seeded clouds to prevent rain from falling over the Chernobyl area. Still others claim that the clouds were seeded to bring radioactivity down on Belarus instead of letting it get to Moscow. Whatever the reason, a huge amount of radioactive rain fell on the country.

Judging which of the three most affected countries suffered the most is not straightforward and the figures are often confusing. For example, the disaster removed 1 million hectares of Ukrainian farmlands and forests from service compared to 464,000 hectares in Belarus By that criterion, Ukraine's economic loss might seem greater. In fact, Belarus suffered the smallest total amount of contaminated land. Russia— which rarely rates as much attention as Belarus and Ukraine—suffered the most, largely in the Bryansk region neighboring Belarus and in the Kaluga-Tula-Orel region 300 miles northeast of the reactor.

Consider that all three countries define land as "contaminated" if cesium-137 levels on a square kilometer of land exceed one curie. But one curie is a relatively low amount of radioactivity when it is spread over such a large area. Natural background radioactivity from radon gas is between one and five curies in many inhabited parts of the world. So some of the "contaminated lands" are, relatively speaking, not really very radioactive.

Nevertheless, this means that there are 43,500 contaminated square kilometers in Belarus, 59,300 in Russia, and 53,500 in Ukraine. Thus, using the one-curie criterion, Belarus suffered the least and Russia suffered the most. But what do such figures really mean given the vast differences in size and contamination levels between the three countries? Belarus's contaminated lands amount to 23 percent of its total territory; Ukraine's, 5 percent; and Russia's, 1.5. Moreover, if 24 percent of all the cesium that Chernobyl released fell on Russia, 20 percent on Ukraine, and another 20 percent on Europe, then fully one-third—33 percent—blanketed Belarus. And because Belarus's contamination was concentrated in a smaller area, radiation levels and exposures were higher.

Yet a different picture emerges when looked at in terms of human suffering. Altogether over the years, Ukraine evacuated and resettled 163,000 people—more than Belarus's 135,000 and far more than Russia's 52,000. This occurred because the parts of Ukraine contaminated with the 15 to 40 or more curies that mandate evacuation were more densely populated than similar regions in Belarus and Russia.

Of course, the flip side of not evacuating people is that they continue to live on contaminated territories. Leading in this category are 1.8 million people in Russia, followed by 1.5 million in Belarus and 1.1 million in Ukraine. There are more Chernobyl thyroid cancer cases in Belarus than in Ukraine and fewest in Russia. But Ukraine contributed the vast majority of the cleanup workers known as liquidators.

So, there really is no correct answer to the question of which country suffered most from Chernobyl. It all depends on what criterion you are using, as shown below. By nearly every measure, though, Belarus surely suffered greatly.

- **Russia**   Greatest amount of contaminated land

- **Ukraine**   Greatest number of people exposed to radiation
    Highest levels of contamination
    Inheritance of the Sarcophagus and radioactive
        waste dumps

- **Belarus**   Greatest percentage of affected land and people
relative to total national territory and
population
Highest percentage of total radionuclides released
Greatest number of thyroid cancer cases

Although the PSRER was a sink for strontium and plutonium, little of which left its borders, this was not true of the light and gaseous cesium that vaporized in the explosion and fire. The reserve's cesium contamination exceeds 1,350 curies in places, but this accounts for only about a third of the radiocesium that fell on Belarus. Much of the rest fell around Gomel and Mogilev, which are surrounded by patches of contamination that exceed 40 curies per square kilometer.

Where cesium density was between 5 and 15 curies, resettlement was voluntary and guaranteed, which meant that people who wanted to leave were supposed to be given free housing. Khoiniki and most of the villages surrounding the radiological reserve were "voluntary, guaranteed-resettlement" villages in which crumbling and empty ruins testify mutely to the people who left. Such incomplete evacuations were the least successful because infrastructure development slows when everyone expects a place to be resettled, adding to the sense of abandonment of those left behind. You're unlikely to find a school in such villages, much less a library, a café, or the "culture buildings" of Soviet times.

The sole exceptions are in larger towns such as Khoiniki, where one can find a clinic (albeit an old, dark, and dank one), schools, a few cafés, and decently stocked shops. But with 5 to 15 curies per square kilometer, Khoiniki is sufficiently radioactive for its residents to have the right to voluntarily resettle. More than 40 percent have taken up the government's offer. Many of those who left were young people—leaving the town with largely empty streets, an elderly population, and demographics skewed towards high morbidity and mortality.

A similar situation is true of Ivankiv, Khoiniki's sister city in Ukraine. Like Khoiniki, Ivankiv is 30 miles from the nuclear reactor, except it is to the south rather than the north. It has a similar elderly population.

In places where contamination was less than five curies per kilometer, resettlement was also voluntary but not guaranteed, and people had to move at their own cost. This may not seem like much in the way of government relief, but in Soviet times people needed government permission to legally change their residence and "voluntary resettlers" were entitled to get it. People who stayed in the one- to five-curie zones, also known as "zones of periodic radiation control," were supposed to receive regular monitoring of contamination levels in the land, food, and water.

Despite the huge resettlements—totaling 350,000 people in Belarus, Russia, and Ukraine by the year 2000—between 100,000 and 200,000 people continue to live on the 15- to 40-curie territories that are officially considered uninhabitable. A 2002 United Nations report maintained that most of them were not at risk from radioactive contamination, although the people would probably disagree.

In one study, people in contaminated Gomel considered themselves less healthy than those in a comparable but uncontaminated town in Russia. But these differences may have been due as much to psychosocial factors as to radiation-induced illnesses, since neither population was particularly healthy.

The greatest danger in the contaminated areas was from consuming homegrown milk and meat. But the people who lived there were less likely to believe that they could do anything about their own health. Fatalists especially tend to have relatively higher doses because they don't even try to reduce them. One overweight woman in her late fifties who suffered from high blood pressure refused to diet or cut her dietary salt because she thought she "will die soon whatever I do."

So, should they stay or should they go? At this point, those that wanted to leave have already done so and of those that haven't, not many want to go. In 1996, Ukrainian sociologists found that 80 percent of the people living in affected areas of Ukraine wanted to leave. Six years later the number had dropped to 20 percent. As a rule, the rural poor who are at highest risk do not want to move.

While the evacuations and resettlements immediately following the accident probably significantly reduced radiation doses, the health benefits of moving people more than a decade after the disaster are not so clear. Some experts say that people received 90 percent of the total Chernobyl-related dose they will get over their lifetime between 1986 and 1995. Resettlement won't change that. Besides, moving people also has costs in terms of both health—the stresses, car accidents, heart attacks—and finances, especially since the same money could do much to improve quality of life in the contaminated regions. Given economic improvements, country people who currently have no choice but to grow and gather their own food could earn the money they need to buy clean produce.

The United Nations recommended that less severely affected areas should be put back into productive use as soon as practicable, and Belarus was evidently taking this advice by encouraging immigration to places that had been evacuated.

One of these places was Strelichevo, a partly evacuated, voluntary, guaranteed-resettlement village that we drove through just outside the northern border of the reserve. Ethnic Russians from former Soviet republics such as Tadzhikistan, Moldova, and Armenia have been settling in the homes of people who have moved away. Most of the new residents are atomically "fresh" and had not been exposed to radiation when it was most dangerous, before 1995.

Rimma asked if I wanted to stop there. The setting sun had finally peeked out from the clouds, reflecting off the wet streets and making the notion of leaving the car rather appealing. But I was unlikely to encounter candor in the presence of a bureaucrat like Palytayev and so declined. Besides, I was tired after traveling the whole day and wanted to see our destination. Expecting (actually hoping) that we were going to be taken to the reserve's headquarters in Khoiniki and given accommodations similar to Chernobylinterinform's hotel, I asked our host where he was taking us.

Flashing a mischievous smile as though he had a secret, Palytayev turned to look at us in the back seat. "I'm taking you deep into the woods and leaving you there," he said, albeit without much real threat in his demeanor.

"Great! I love camping," I said but wondered what he had in mind.

What he had in mind was the evacuated village of Babchyn. The Polissia State Radiological Ecological Reserve's headquarters was in Khoiniki, but its base inside the reserve was in Babchyn, making it the functional equivalent of the town of Chornobyl in the Ukrainian zone.

Before we passed through the checkpoint guarding entry to the village, Palytayev handed me a temporary permit to enter the reserve. It was dark red cardboard, making it rather difficult to read, but it just had my name and the relevant dates on it. It was quite different from the Chernobylinterinform program, which was a sheet of paper listing all the places you planned to visit, the people you planned to see, and the checkpoints you'd be crossing.

After we drove into Babchyn, our host explained that the dozen or so small houses of white brick bordering the road were built in the early 1980s, just a few years before the disaster. Now they housed the reserve's scientific staff, laboratories, and administrative buildings. It was evening when we arrived and nearly everyone had gone home to Khoiniki except for the checkpoint guards, firemen, and some scientists.

It was a tiny village. Driving through it took less than five minutes. After the firehouse with its lookout tower in a feral meadow sprouting thick grasses and clumps of shrubs, we arrived at a single-story building surrounded by a profusion of well-tended flowers. Painted in pink, mauve, and brick—appropriately, the same colors symbolizing high contamination on the radiation maps—the sign on the building's front lawn identified it as the Sanitary Processing and Dosimetric Control building.

Palytayev really hadn't been kidding. He was about to leave us in the wild—or at least in a practically abandoned village surrounded by not much of anything. But it was peaceful and pretty outside. The rain had stopped and the air was heavy and green. The only sounds under the humid and hazy skies were bird songs and the buzz of an occasional mosquito. My dosimeter measured 60 microroentgens an hour—or about four times natural background—at the one place I tried measuring it in the field across the road. But the radioactive contamination was patchy all over the affected territories. Had I walked just a few feet away, I would have registered completely different levels.

Our host was gallantly helping us carry our things inside, so I followed him around the foot bath cemented inside the front vestibule for washing radioactive dust off shoes during Chernobyl's early aftermath. He gave us a tour of the accommodations, which were just fine, if a bit run down. The locker-room-style showers, without walls or doors, were for the reserve's workers coming back from highly contaminated areas. The hot water got shut off at 7:30 p.m., when the spectacled fellow in charge of the boiler and pumps caught the last shuttle bus to Khoiniki. There was also a refrigerator, but so much ice and snow had built up in the freezer from frequent power outages that it barely worked.

It was good that I had brought food. The refrigerator was empty but for a loaf of black bread and a jar of homemade pickles, and there were no stores in Babchyn where we could buy anything. A canteen served lunch, but the cooking staff had left at 4:30.

I unpacked quickly in what was designated as my bedroom, a spacious room of brown velour furniture, a pine writing table, and a closet. My dosimeter beeped a reading of 32 microroentgens an hour—almost three times as high as in the Chernobylinterinform hotel in Kiev. In fact, I was surprised at how high it was.

When I emerged, I found Rimma and Palytayev sitting at a long table in the building's conference room. After consulting with me, she had invited Palytayev to join us for a modest meal of bread, cheese, peaches, and a bottle of Transcarpathian brandy that we consumed under the glassy eyes of a boar trophy shot in the reserve.

Although I decided to put my notebook away and relax, at one point Palytayev said something about the Chernobyl disaster that moved me to write in a less sober but still legible hand:

"They thought that it was a peaceful atom. But it was savage."

SANCTUARY

Palytayev picked us up the next morning in a boxy purple jeep known as a UAZ, which were the initials of the once-Soviet automotive factory that built it. It had high clearance and a tight suspension that proved well suited for the narrow roads and trails cross-hatching the radiological reserve.

"The Chernobyl station is 60 kilometers due south," he said after we piled in and started off in that direction on a paved road that was wide enough for just one car. This proved not to be a problem because we didn't encounter even one other car that day.

Passing former hayfields blanketed with wildflowers under cloudy and cool skies, Palytayev explained that 60 percent of the reserve's forests are deciduous. The rest are conifers. Opposite proportions were found in the Ukrainian zone, where most of the trees are pines, which grow well in the sandy local soils. But the Belarusian forestry service preferred to plant deciduous trees after concluding that pine forests were too dangerous because of their flammable carpets of dried needles. There were noticeably fewer pines than in the Ukrainian zone.

After four miles we came upon a checkpoint. Unlike Ukraine,

which controls its 30-kilometer zone's perimeter, Belarus doesn't have costly checkpoints on the reserve's borders—only keep-out signs and trefoil-painted boulders that are too heavy and worthless for anyone to steal. But Palytayev explained that they do have 11 checkpoints inside the reserve at intersections and anywhere that there are more than two directions in which an intruder could drive.

Manned by camouflaged guards working 24-hour shifts, the checkpoint included a well-kept cottage surrounded by flowers. I had never seen anything resembling a garden at Ukrainian Chernobyl checkpoints, which are generally eyesores. Yet every single checkpoint we saw in the Belarusian zone had a nice little cottage and gardens. Beautification was clearly official policy.

All the checkpoints displayed boulders with warnings against hunting and painted with stylized moose, lynx, and roebucks. The paintings were so primitive as to approach naïve and I quite liked them. Palytayev explained that the reserve's employees did them.

He barked some orders into the jeep's radio and the guards ran out to open the gates as we approached so that we didn't have to stop as we passed through. Driving with the boss had its benefits. Soon after the checkpoint, Palytayev made a right turn onto a dirt road and drove west. To our left was the barbed wire fence the Soviets put up in 1986 to mark the 30-kilometer zone. But it was incomplete and didn't extend around the zone's northeastern arc.

We were due north of the reactor. On radiation maps, the territory inside the barbed wire was painted in dark shades of red to indicate contamination levels from 100 to 1,000 curies. Three blotches of brown located less than 10 miles from the reactor marked extremely contaminated hot spots of cesium measuring more than 1,000 curies per square kilometer. They were traces from the original explosion and also contained much of the plutonium that had fallen on Belarus. One of them was right on the Belarusian-Ukrainian border.

On our right was a canal that had once been used to drain the peat lands, called mires, that had dominated the Polissian landscape north of the Pripyat River until they were drained for farmland. My makeshift topographical map of the Belarusian zone was thick with canals.

I showed it to Palytayev and asked where I could buy a map that contained the 15 square kilometers of Belarus that I was missing.

He appeared to admire my handiwork but shrugged. "I have no idea."

As Rimma told me later, never trust an ex-Communist function-ary. In fact, the maps were available in larger towns like Gomel, but I only learned that several weeks and nearly $70 later. Too distressed by the hole in my map after returning from Belarus, I ended up buying the missing piece from the Minneapolis-based East View Cartographic company for $63—and that was with a 20 percent discount they gave me for plugging them in this book.

"There's a black grouse!" Palytayev exclaimed and pointed over his shoulder at a plump bird that flew up out of a copse of birches. I turned to look but caught only a black flash as it flew away. This meant it was a male because the female is a speckled brown. Although I can, with the help of a handbook, often identify the more dramatically plumed male game birds—such as the black grouse with its red eyebrows—most of the speckled brown females fall into a catchall category of "field chicken" that I don't even try to identify without the help of an expert birder.

Polissia is right on the southern border of the black grouse's tradi-tional European nesting grounds, which used to be all over Europe but are now mostly limited to the northern regions of Russia and Scandinavia. But the depopulated Chernobyl lands have created invit-ing environments for many birds that had never been found there since anyone started keeping track of such things. Unfortunately, the Chernobyl people don't talk to the people who write the birding hand-books. There is, in general, very little literature about zone wildlife.

For example, two of my European birding manuals show great white egrets nesting only in parts of Crimea and southern Ukraine. In the Ukrainian birding manual, they nest nearly all over the country except Chernobyl lands and Polissia. In Belarus there were no great white egrets except for the occasional vagrant. But in 1997 the first great white egret egg was found. Since then the zone has become one of the birds' primary nesting sites—especially in the wetlands north of the Pripyat River.

Aquatic birds such as cormorants began appearing in 1988. Twelve years later, there were thousands, nesting in giant colonies in trees that they eventually kill with their accumulated droppings. Stocky, croak-ing night herons were found breeding for the first time in 1999. Blue and white azure tits, rare just about everywhere in Europe, have begun nesting there, as have aquatic warblers—one of the world's rarest birds. Endangered white-tailed eagles have also appeared and some even win-ter in the zone, stealing fish from minks that catch them in ice holes.

Short-eared owls build their nests in fen mires—peat marshes that have been drained into disappearance in most places in Europe. Kestrels nest in the abandoned flower boxes on Pripyat balconies, and young endangered eagle owls can be spotted near the Sarcophagus. Although the figures vary depending on what different birding experts consider a "sighting," from 250 to 280 bird species—40 of them rare or endangered—have been spotted in the zone since the evacuation.

Moreover, the zone lies on a major migratory path. Twice a year, about half a million birds pass through it, consuming radioactive food and carrying radionuclides to their destination countries. Eight years after the disaster, song thrushes that wintered in Spain had detectable amounts of cesium-137 and strontium-90 in their muscle and bone. These weren't dangerous levels, but the birds were from central and northern Europe, places that were far less contaminated than the zone.

Because it mimics calcium, radioactive strontium concentrates in eggshells, where it bombards vulnerable, growing embryos with beta particles. The eggshells of great tits nesting in the highly contaminated Red Forest contain as much as 40,000 becquerels per *gram*—an extraordinarily high density of radioactivity that is comparable to solid nuclear waste. While normal birds' eggs are uniform for a given species, Red Forest great tits' eggs vary greatly in their size and shape even within the same nest. In a 2003 study, many eggs were empty or contained dead embryos, and from the eggs that hatched, fewer birds ended up leaving the nest compared to control birds. In one-quarter of the nests, none of the nestlings survived, although the causes of death were not always clear. It could be that their parents were weakened by the effects of radiation and didn't have the energy to feed their young. Great tits in the Red Forest display pathological changes in their blood.

Soviet-era research on nuclear spills in the military complex found that strontium-laced eggs also contained deformed embryos.

But the only avian mutants identified in Chernobyl lands have been barn swallows with partially albino faces instead of the species' typically rust red chins. It's possible that the birds had high radiation exposures in the early period after the disaster because they flew low over the fallout-coated fields, and the metabolism of carotenoids, the organic molecules responsible for their red plumage, is especially susceptible to radiation. Whatever the cause, the albino-speckled males are evidently not very attractive and fewer females choose them as mates.

Soon we drove into the village of Tulgovichi. It overlooked the Pripyat River at a place that appeared to be right outside the reserve on my map, about 30 miles northwest of the nuclear power plant. The map marked it as occupied, though this was hard to tell as we drove past rotting cottages with collapsed roofs, overgrown yards, and blank windows. Weeds and shrubs grew out of long abandoned white storks' nests.

"We've tried attracting white storks by putting wagon wheels and other platforms on poles for them to build nests," Palytayev explained as we trundled slowly through the village. "But they don't come here. They like places where there are people cultivating fields."

Tulgovichi had 12 families but none of them were doing much in the way of cultivation, and none of them were evident on the streets either. There were plenty of pigs, though. Aside from a sow and half a dozen piglets snorting about the roadside, several others were just wandering around at a crossroads where radiation levels on the grassy shoulder I measured at random were 65 microroentgens an hour. This was high—higher than nearly every inhabited place in the Ukrainian part of the zone.

We stopped in to visit Nikolai Shamenko, who lived in the first occupied house we came across. It was a filthy, unkempt place where 80-year-old Shamenko was sharing a smoke with his 45-year-old son, also named Nikolai, who lived across the street.

"We live well," insisted the older Nikolai without prompting after the three of us crowded awkwardly into the front room of his traditional two-room cottage. One of things I dislike most about journalism was playing tourist in other people's lives, especially when those lives were so difficult compared to mine that I felt stupid and intrusive even asking them questions.

Luckily, old Nikolai was voluble. "Actually, we live badly," he admitted and explained that he had never been evacuated but that his son Nikolai had.

That prompted the younger Nikolai, who had been lying on a pile of rags and smoking, to join the conversation. His face was shiny and his grimy pants were belted with a length of twisted pink fabric.

"They gave me an apartment in Mozyr," he said, referring to a town about 10 miles away on the Pripyat River. "But I gave it to my kids, left my wife, and moved back here with my new wife. We live across the road."

Palytayev sat on a chair near old Nikolai, who grunted at him and asked: "Who are you?"

"I'm the director of the reserve," Palytayev responded.

Old Nikolai squinted skeptically: "What reserve?"

"The radiological reserve where you're living."

Old Nikolai seemed even more skeptical and pointed at Palytayev's paunch. "Our director is thin, and you're fat!"

I barely stifled a laugh. Clearly, Palytayev—who had started his job only a year earlier—did not go visiting very much in the reserve. And he didn't want to stay with the Shamenkos very long. This was fine with me.

After saying our good-byes, we piled back into the jeep and drove past the obviously occupied white brick cottage that was Nikolai-the-younger's home. It was neat and surrounded by flowers. I didn't want to make sexist assumptions that these were his wife's handiwork. But Nikolai did not look as though he cared much about appearances, leading me to wonder about the woman who had agreed to set up house-keeping in a place as radioactive as Tulgovichi.

Palytayev drove off the paved road into a maze of dirt paths through the largely empty village before emerging onto a short strip of asphalt that ended on a cliff overlooking the Pripyat River.

"The hydrofoil between Kiev and Mozyr used to stop here every hour," said Palytayev when we piled out of the jeep. "That's why there's asphalt here. It used to be a pier."

All that remained of the pier were some concrete blocks that had collapsed at the bottom of the cliff.

"It was also a popular place for camping," he said, gazing at the riverbanks that had once been crowded with tents. A pair of gray herons stood motionless in the water.

"And then, in one moment, it all ended."

After a filling four-course lunch in Babchyn's canteen, Palytayev drove us out to neighboring Vorotets, a village that had once boasted a swine farm of more than 20,000 pigs and now housed the reserve's experimental farm. In contrast to the gradual destruction of human habitats that we saw on the way there—like the trees breaking through the rotting roofs of wooden cottages, darkened by time—the farm buildings were freshly whitewashed. But the paddocks and stockyards were empty.

Palytayev parked the car near a barn and we climbed out. The sun had come out and the weather suddenly turned humid and uncomfortably hot.

"Where are the mustangs?" he asked some men wearing coveralls and repairing a fence. He was kidding about the mustangs. The farm raised two breeds of Russian horses, a heavy draft as well as a lighter trotter, and had been doing so quite profitably for eight years. Each year, 30 horses were sold for about $500 each, mostly to private farmers. When we came to visit there were about 150 horses in the herd—only they were nowhere to be seen.

"People don't eat horse meat, so the fact that the horses have internal contamination doesn't matter much," he said, explaining that in the summer the horses graze outdoors on radioactive grass while in the winter they are fed hay and oats, also grown in the zone and also radioactive. But Vorotets, like Babchyn, wasn't highly contaminated and the horses weren't either.

The farm had 250 pigs that produced piglets for sale. Sold for a nominal price (much more cheaply than the horses) to local rural folks who fattened them on uncontaminated feed, the pigs' meat was supposed to be clean by the time they were slaughtered.

Aside from stockbreeding, the administration also had a lumberyard that made pine furniture. All of the checkpoint cottages were decorated with pine desks, hutches, and benches, as was an entire wall of Palytayev's office, which was built to resemble the wall of a log cabin.

The Belarusian reserve's businesses stood in marked contrast with Ukraine, which banned any kind of economic activity in its zone except that connected with cleanup activities and providing services to the zone administration and workers. I had once asked the folks at Chernobylinterinform why they didn't sell tasteful T-shirts and coffee mugs to raise a little money for their black hole of a budget, only to learn that it was forbidden.

Or so they said. Maybe it was simply that no one there wanted to go to the trouble of organizing production.

Vorotets's experimental farm also had 33 dairy cows that happened to come in from their pastures just when we arrived. Personally, I find cows to be among the scariest of domestic animals, and I stood frozen as they clattered around me on their way to one of the barns.

Soon afterwards, a young blond horseman cantered up to us on a sweaty gelding covered with swollen horsefly bites. All that remained of the saddle were the frame and some patches of leather, while the bridle was homemade and made of braided hemp rope. But the tack evidently worked well enough for the rider to canter off at Palytayev's command to find the horses.

When he returned with news on where they were grazing, Palytayev herded us back into the jeep and drove back to Babchyn, but on a different road that took us past fields that had been plowed to plant grassy pastures for the horses and the construction site for the reserve's new canteen and administrative buildings.

Ukraine doesn't allow any plowing or building in the zone because it raises radioactive dust. With the vast majority of radionuclides in the upper 10 centimeters of soil, plowing and building expose the most contaminated layers to the air.

Palytayev seemed unconcerned about the risk and more interested in the fact that all of the scientists and administrators would be under one roof—in a new building with its own boiler and generator.

The horses were in a large field just behind the site. Nearly all of them were a light chestnut color, but none resembled heavy Russian draft horses because someone let the draft horses mix with the trotters. It didn't really matter, though. Belarus was too poor to support much of a sport horse or racing industry, so there was not much demand for purebred trotters. The mixed breeds were perfectly adequate plow and cart horses for private farmers with small plots of land. They were also used by the reserve's fire and security patrols. The reserve was huge and the Belarusian budget was too poor to provide much money for gasoline—especially after world prices shot up in the wake of the U.S. invasion of Iraq.

In fact, after showing us the horses, Palytayev stopped by the Babchyn fire station where one of the firemen poured a canister of gasoline into the jeep. The gas-guzzling vehicle had used up a full tank during our travels that day, and he needed enough to get back to Khoiniki. Since we wouldn't see him the next day and would instead get a ride back to Chornobyl from one of his deputies, it was also time to find out what, exactly, the visit would cost me.

At the equivalent of about $100—much of it for gasoline—it turned out to be much cheaper than I expected. After I gave him the money, Palytayev drove us to a laboratory housed in one of Babchyn's brick cottages and put us in the care of Yuri Bondar, a radiochemist who was the reserve's deputy science director.

The 4:30 bus had already left for Khoiniki and Babchyn was almost empty.

"It's so lonely here," Rimma said, looking up and down the streets. "In Chornobyl, there are a lot of people working on a shift, even at night. Here, there's nobody."

The nightlife in Babchyn was strictly BYOB.

Lanky, balding, and deeply tanned, Yuri Bondar had just arrived that morning to begin his two-week tour of duty at the reserve. In contrast to Ukraine, where many zone employees worked tours of duty, in Belarus only highly qualified scientists did.

"That's because there are no scientists in small towns like Khoiniki, so all of us come down from Minsk," said Bondar, who had been working in the Belarusian nuclear energy institute in 1986 and remembered arriving at work to find all the radiation meters going inexplicably crazy as the radioactive cloud passed through Minsk, about 200 miles to the north.

Minsk's distance from the reserve is one reason most outsiders perceive Chernobyl as a largely Ukrainian problem. Chernobyl is merely a two-hour drive from Kiev. But getting there from the Belarusian capital takes five hours and that's on a good day. Moreover, because of the complicated border crossing, foreign journalists who want to see the Sarcophagus and the ghost city of Pripyat—and most do—usually opt to see the Ukrainian zone. That's why descriptions of Chernobyl, such as its distance from large cities, usually have a Ukrainian context.

"We didn't know anything about Chernobyl at the time," Bondar continued. "But when I saw the radiation levels, I said that it was worse than an atom bomb, and unfortunately, I was right."

The wind picked up as a dark and imposing storm cloud on the horizon headed our way to the accompaniment of alarming thunder and lightning. I generally enjoy thunderstorms if I am safely indoors, but there was something about Babchyn's isolation that made me feel less than safe.

We had invited Bondar to join us for dinner and drinks back at the Sanitary Processing and Dosimetric Control building, but since none of us had umbrellas, we decided to wait out the storm in a dark laboratory crowded with computers and radiation meters.

The room was dark because the electricity had been turned off during the storm. Bondar explained that this was done in villages all the time as a precaution against power surges, but it was a big head-ache for the radiological reserve's staff in Babchyn. Although the lab's computers and spectrometers had small generators that kicked in au-tomatically when the power went off to prevent data losses, other zone services were more affected. For example, the boiler that heated the water in our building was unaffected by the power outage, but the pumps that got the water from the boiler to the shower were electric. So, if reserve employees came back from doing some radiologically dirty work during a storm, they wouldn't be able to take showers.

The phones were also a problem. The Soviet-era village lines made it nearly impossible to get a good Internet connection. Bondar some-times had to spend hours trying.

"The phone lines in Chornobyl are terrible, too," said Rimma.

No wonder Palytayev was so happy about the new administrative building, even if its construction did raise some radioactive dust.

Luckily, the storm blew over quickly, with hardly any rain. The pavement was mostly dry when we walked the 10 minutes it took to get from the cottages to our building, where Rimma and I quickly set out a picnic on the stoop. Potato salad with sun-dried tomatoes and bal-samic vinaigrette, lightly curried chicken with bacon.

Bondar did the honors of pouring the Transcarpathian brandy. Traditionally, only men were supposed to touch open bottles of alcohol.

I asked him about the rumors that Belarus was so highly contami-nated because cloud seeding brought radioactive rain down on the re-public to keep it from hitting Moscow. Judging by his expression, he had heard the rumor before.

"It's entirely possible," he said. "But if it did happen, it was done in secret and no one will ever admit it."

Bondar seemed more open to talking about the subject than other Belarus scientists, who fell into uncomfortable silences when I raised it with them at Chernobyl seminars and conferences. Maybe it was the brandy.

But in fact, all major cities and towns in the region were relatively unscathed by the disaster, and rumors persisted that clouds had been seeded to prevent rain from falling on densely populated urban centers.

"Look at Gomel," said Bondar. "It's relatively clean, too."

The regional capital of the most contaminated territories outside the borders of the zone, Gomel was 80 miles from the nuclear plant. But although it was surrounded by dark colors on the radiation maps, signifying cesium contamination levels of 15 to 40 curies or more, Gomel itself was a moderate 1 to 5 curies per square kilometer. The same was true of Mogilev, 180 miles to the northeast in Belarus; Chernihiv, 40 miles to the east in Ukraine; and Kiev, 50 miles to the south. All of these cities were surrounded by moderate, if notable, radioactivity levels but were relatively clean themselves. Even Pripyat lay outside the lobes of lethal contamination from the initial explosion.

Bondar thought it might have to do with the effects of so-called urban heat islands, a phrase reflecting the fact that urban areas are warmer than their surrounding suburbs and rural areas. Buildings, roads, and other artificial surfaces retain heat; cars and air conditioners produce it, and the higher temperatures that result can, for example, increase summer rainfall levels downwind of cities. Maybe urban heat islands played a role in ensuring that the clouds didn't rain radioactivity on urban centers.

But don't tell that to the old babushkas in Chernihiv. They say that their city was saved because of the eleventh-century monastery that was built to mark the spot where a miraculous icon was found on the banks of the Desna River. It has, according to local lore, protected the city ever since.

FLOURISHING IN THE FENS

The next morning Mikhail Rupashchenko showed up at the Sanitary Processing and Dosimetric Control building to collect Rimma and me in an old Soviet van. He was Palytayev's deputy, a 30-something forestry professional responsible for the eastern quadrant of the radiological reserve, the part we'd be driving through to get back to the Paryshiv checkpoint in the Ukrainian zone.

"It's an interesting job," he said. "But it's not like I have much choice."

Although it was called the "Bragin Sector," after the decontaminated district center 30 miles northeast of Chernobyl, Rupashchenko's division was headquartered in Komarin, just a few miles from the Belarus-Ukrainian border crossing. It was the type of town where you were lucky to have any job, even working in a radioactive nature reserve.

"Komarin is 29 kilometers from the plant, which placed it inside the limits of the original 30-kilometer zone," Rupashchenko explained, and when I examined my ersatz map, I was surprised to find this to be true. In Ukraine the barbed wire borders and checkpoints are such a significant part of the zone's imagery that it was hard to accept that parts of Belarus's 30-kilometer zone were inhabited and openly accessible without permits. On the maps of cesium contamination, Komarin was in the one- to five-curies range.

"Komarin provides better access to the reserve's Bragin Sector than Bragin itself," said Rupashchenko, who had blue eyes, a gold incisor, and a penchant for telling jokes that got increasingly off-color as the day progressed. Many had to do with the high sex drive of Belarusian women, a subject about which I knew nothing and so couldn't tell how funny the jokes were. Rimma laughed, but I only smiled politely at what seemed to be the punch lines.

We were driving south from Babchyn, on the same road towards Chernobyl that Palytayev drove on the previous day. But instead of going west, Rupashchenko turned east onto a packed dirt trail that ran alongside a drainage canal, taking us past ghost villages gradually being consumed by the wild. The only fence in the reserve was the barbed wire perimeter of the 30-kilometer zone. We drove along it for a while, but then it ended.

"They just stopped building it in 1986," said Rupashchenko, who didn't know why.

Abandoned barns and silos were rotting away jigsaw patches of wood, paint, and metal, and the remains of cottages were barely visible through the thicket of vegetation.

"See that grass," said Rupashchenko, sweeping his arm outside the van's window.

It was hard not to see the grass. There was practically nothing to see *but* the grass.

"That's at least two meters high," he continued. "If you went in there, you'd disappear from sight. If a moose was in there, you might just see the tips of its antlers. A herd of boars could be in there and you wouldn't see them at all." Actually, Rupashchenko used the Slavic word *los'* to refer to the large, roman-nosed deer called "moose" in America and "elk" in Europe. Confusingly, in North America, the word "elk" refers to red deer.

After lying fallow for 18 years, the drained swamps were blanketed with tall grasses and about two feet of thatch that was like gunpowder. One match would spark huge radioactive brushfires.

After about half an hour, we were back on a paved road that took us past Savichi, an inhabited village like Tulgovichi that was right on the reserve's borders. This took us outside the reserve for a few minutes, but the road passed back inside again, taking us through a checkpoint guarded by camouflaged men in yet another white brick cottage surrounded by flowers. I had hoped to use their outhouse, but it was flooded after the previous days of rain, so I ended up going in the woods where a horsefly bit me painfully, leaving a mark on my thigh for months.

Back on the road, Rupashchenko explained that we were on our way to a place called the Upper Swamps, where Belarus had started restoring, or at least rehabilitating, the drained peat lands around the villages of Kriuky and Kulazhyn. On the radiation maps, there are two big brown blotches between them, representing cesium-137 levels in excess of 1,000 curies per square kilometer.

In the very hot and dry summer of 2002, huge fires ignited in the peat lands around Moscow and in Belarus. The smog suffocated the Russian capital and drifted as far as Sweden. Kiev, too, was blanketed in a strange smoky haze from peat fires that ignited north of the capital, increasing the cesium levels in the air. Because the soil itself is literally aflame, peat fires can burn deep downwards and last for weeks. Luckily, the peat lands around Kriuky and Kulazhyn were not among them, but the prospect that they could catch fire and release their massive inventories of radiocesium into the air prompted the decision to flood them.

To flood the swamps, the reserve had simply dammed the ditches that had been used to drain them. It was a cheap and effective system that had been used to reflood 4,000 hectares, or about 15 square miles, over the course of two years. The dams required maintenance, though. If they broke, the swamps would drain again.

"Filling up the drainage ditches would be the best solution," said Rupashchenko, whose sector contained most of the mires. "But that's expensive. We've achieved what we wanted. There haven't been any peat fires since 2002."

Experts argue whether reflooding peat lands really counts as restoration, or merely their rehabilitation, since "restoration" implies recreating wetlands that are the same as they were before they were drained. Over the short term, this is impossible to do, although it may happen over many years. In any case, Rupashchenko didn't use either word but instead used the term "reflooding."

It was not yet clear whether the reflooded peat land would become a bog or a mire. Although the dominant vegetation consisted of reeds, making it more like a fen, only two years had passed. Rupashchenko predicted that it would look entirely different in five or six years.

Whatever it was called, it had created a paradise for birds. As we approached the Upper Swamps, black storks glided about the car, their bright red bills piercing the air like daggers.

"Black storks are becoming so common, they are no longer endangered in the reserve," Rupashchenko told us. "But white storks are getting much rarer."

Although one reason was probably the lack of people and the absence of cultivated fields full of frogs and snakes, he thought it might also be because the number of raptors had grown markedly and baby storks exposed in their high and open nests were vulnerable to predation. The shier and rarer black storks conceal their nests in trees.

After spotting at least half a dozen black storks, I pleaded to stop the car for a closer look.

But Rupashchenko laughed and said: "This is nothing. There are many more of them up ahead."

It was true. Gray herons, mute swans that took off at our arrival like commercials for KLM, and thousands of ducks that rose into the air in a tornado-like cloud shared the flooded peat lands with dozens and dozens *and dozens* of great white egrets. There were so many egrets that I could only begin to count them before our appearance made them take off deeper into the renewed wetlands, their flight seemingly silent at a distance that made them look like fleeing wraiths.

"All of the reflooded peat lands have become bird sanctuaries just like this one," said Rupashchenko. "If you come here in the morning or evening, the birds make such a racket you wouldn't be able to hear me talk."

"It's so beautiful," I said, gazing through my binoculars.

"If only it wasn't radioactive," responded Rimma.

"If it wasn't radioactive, it would be a farm—and there wouldn't be any egrets," I said. It was one of Chernobyl's more profound ironies that never failed to affect me and drew me back again and again like a magnet.

When I placed my dosimeter on the thick grass by the roadside, it measured 250 microroentgens an hour. At waist level, it was 200. And this was 10 miles from the nuclear power plant. On the Ukrainian side of the border, there were no places with such high radiation levels so far away from the reactor. Although the Upper Swamps were mostly contaminated with cesium, and birds' eggshells were not as packed with strontium as those in the Red Forest, it was nevertheless a very radioactive environment. But all indications were that the radiation was benefiting the birds because it had gotten rid of the people. Whatever the effects of cesium, strontium, and plutonium might be on individual egrets, they were not as bad for egret populations as human activities.

The downside of flooding the peat lands was that radionuclides are very mobile in bogs and fens and transfer much more easily into plants than they do in drained peat lands, although they can be very mobile in those as well. As with all things Chernobyl, it was a choice between lesser evils. The bitter wormwood waters and plants of the renewed mires were less dangerous than the risk of peat fires.

We were also about four miles from Kriuky and the brown patches of cesium on the map, where radiation exposure goes up to three milliroentgens an hour. Ours was a rose-colored patch.

Back in the van, Rupashchenko drove us a short distance into the forest to show us three rare black birch trees. Although they occur occasionally in Polissia, black birches don't appear in handbooks of European trees, and botanists differ over whether they are a distinct species or merely variants of white birches.

But Rimma was more interested in the rich harvest of champagne-colored chanterelle mushrooms that sprouted in the birches' shadow after the previous day's rain. My dosimeter beeped a rapid 250 microroentgens an hour on the ground. It really was a hot patch, and the mushrooms were probably loaded with tens of thousands of becquerels if not more.

For obvious reasons, mushroom picking is forbidden in the zone, though that didn't stop Rimma from bemoaning the waste of the lovely looking chanterelles. Mushroom hunting is a traditional pastime that has become a survival mechanism during the post-Soviet economic downturn. I didn't even want to touch them.

A bull moose watched us from the other side of the road, standing motionless between some bushes about 50 yards away. His gray-brown pelage camouflaged him perfectly, and I wouldn't have seen him at all had Rupashchenko not spotted him first. His eight-point antlers were still covered in velvet, but he would fray it in time for the autumn rut, when the radioactive reserve would echo with the moose's nasal, squeaking mating calls. After obligingly letting us watch him for 10 minutes, the moose trotted off with a clumsy, ambling gait until we could no longer see him through the thick vegetation.

I had never seen a moose in the wild, though they had always held a fascination for me after they began returning to the northern reaches of my native New York State in the 1980s. After a decade of living in Ukraine and spending time in a number of natural landscapes from Crimea to the Carpathians, I had never seen a wild animal larger than a rabbit. Although the moose was standing on Belarusian territory, the depopulated zone on both sides of the border had become an inviting habitat for wild animals—but not only because the newly feral landscape provides room for ranging and foraging: The rules that prohibit hunting for mushrooms also prohibit hunting for game.

4

Nuclear Sanctuary

*I will let loose wild animals against you and they shall bereave
you of your children and destroy your livestock; they shall make
you few in number, and your roads shall be deserted.*

Leviticus 26:22

Rats and roaches—in fiction, movies, and Internet chat sites,
these familiar pests are ubiquitous parts of imagined
postnuclear landscapes, their adaptability and fecundity supposedly giving them the ability that humans and more cuddly creatures lack, to survive in radioactive wastelands.

Although the conventional explosion at Chernobyl didn't have the
blast of a nuclear bomb, much less the unimaginable devastation of
all-out nuclear war, the dread permeating the most devastating por-
traits of doomsday are less connected with the immediate effects of a
nuclear exchange. Indeed, most nuclear war fiction has devoted little
attention to the fireballs of fusion blasts and has focused more on the
fate of survivors in the deadly irradiated environment or on the lives of
some distant descendants, for whom the nuclear war is ancient history
but the effects of its lingering radiation continue to devastate.

Both reflecting and fueling popular imagination, these postnuclear
badlands are often populated by fantastic and repellant mutants—both
human and not—some enhanced with superpowers such as telepathy
or grown to gigantic size. But even rodents and insects of standard size
crawl across the radioactive rubble in the doomsday visions of scientific minds, like that of Stanford University biologist Paul R. Ehrlich,

who once gloomily predicted a postapocalyptic world: "There may be a few [human] survivors in very deep, very well-stocked shelters, but there will be nothing for them to do when they come out. They'll mostly serve as food for cockroaches and rats that are likely to survive the war much better than human beings."

As if to confirm all of the darkest scenarios, rodents actually did have a population boom after Chernobyl. In 1987 and 1988, house and field mice seemed poised to overrun the evacuated zone when their numbers exploded from about 20 to 30 per hectare to as many as 2,500! Evidently attracted by plentiful food in the unharvested fields left behind after the evacuation, the rodent problem became so acute that some zone authorities wanted to poison them. But biologists stepped in and predicted that the population would soon stabilize on its own. And that is exactly what happened.

First the population explosion attracted predators: foxes, weasels, and especially raptors. In just one square mile of meadow near the buried village of Kopachi in the 10-kilometer zone, there were enough rodents to support marsh harriers and short-eared owls, kestrels, and falcons.

Still, there were too many mice and there wasn't enough in the fields for all of them to eat. But these critters have small ranges and couldn't go on long treks in search of food. Nor could they escape into the neighboring forest to which they are not adapted. So, in the autumn of 1988, most of the mice starved. This, in turn, caused another temporary boom in the number of meat-eating scavengers that descended on the bonanza of dead. But once the fields were cleansed of rodent corpses, nature's sanitation workers also left. It was one of the first examples of how, in the absence of human intervention, nature in the zone could recover its balance.

It is a balance that now includes radiation, though even this is a mutable quantity that depends on the multitude of factors affecting any ecosystem. Radionuclides migrate with water and wind. Levels of contamination in animals have varied over the years as some radionuclides decayed and others made their way through the food chain, often in unpredictable ways. Thus, although the amount of radionuclides in animals decreased steadily in the first three years after the disaster, it started rising again in the fourth year with a peak in 1992-1994. This anomaly has yet to be fully explained. One theory is that the pine needles that bore the brunt of the initial fallout took six to eight years to decompose fully. So they began adding their considerable stores of radionuclides to the food chain only in 1992. But this theory doesn't explain why moose shot in Belarus in 1992 had the lowest levels of radioactivity of the postdisaster years—though it could be because Belarus shoots few zone animals for science annually, so the samples aren't statistically representative.

In fact, very little is known about the radioactive animals of Chernobyl.

What is known is that there are many, many more of them than before the disaster.

PLUTONIUM SAFARI

The zone in late autumn was a subtle landscape, painted with a cool palette of green pines, pale yellow fields, and silvery birches capped with a filigree of copper branches. Bare willows provided unexpectedly bright splashes of orange that framed the road on an unusually sunny and warm November afternoon.

I was in an old green Soviet army jeep that spouted exhaust fumes as it carried my companions and me over a narrow, potholed road sprouting shrubs and scattered with moss rugs. We were in a deep-orange part of the radiation maps, where the 1986 winds dumped 20 to 50 curies of cesium-137 on every square kilometer.

After passing a battered white sign for Zapillya—about five miles west of Chornobyl—the driver turned left onto a sandy trail. It was a fire line that was supposed to lead us to what everyone hoped would be glimpses of the zone's big game. Actually, I tried to keep my hopes low. Large animals are generally shy. Despite many trips to the zone, the only large animal I had seen was the moose in Belarus. On the Ukrainian side of the border, the largest wild mammal I had seen was a fox pouncing in a field. But at the very least, I thought I might learn a little about tracking.

My guide that day was Oleksandr Berovsky, whom everyone called Sasha. A strapping 29-year-old with buzz-cut blond hair, Sasha was captain of the Chornobyl fire department. But his passion was animals. A veterinarian by training who became a fireman only as an alternative to the military draft, Sasha satisfied his original calling by raising pheasants, quails, ducks, and rabbits in the fire department's outbuildings and spending his free time riding the department's horses in the zone's wild lands.

Yuri Kolesnik, another fireman with a neat black mustache, often joined him on those trail rides and had joined us on our wildlife expedition.

We were following the jeep tracks that Sasha had left the previous day, when he spotted a lot of wildlife. Since most wild animals don't wander aimlessly around their territory but follow a network of paths that they know intimately and can use to escape if necessary, it made sense for us to follow his tracks. Soon enough, we left the smooth sand and turned onto a narrow, rutted path through brush and forest.

"You can only do this ride in a UAZ," said Sasha after the jeep bounced its way over a rough patch through a thick jumble of branches that slapped the windows. It was similar to the jeep that Palytayev drove in Belarus, but in worse condition.

"Or an SUV," I said.

"It would get all scratched from the overgrowth. That would be a shame. With the UAZ, it doesn't matter," said Sasha, who had offered the UAZ after hearing that I intended to drive my 1998 Nissan on the expedition. Actually, I thought that we were going to drive into the woods and then go on foot to look for tracks. Evidently, that was not Sasha's plan, at least not the way I imagined—and it was a good thing, too.

"This was once a rye farm," Sasha explained when we came upon a field of tall grasses and bushes. Then he jerked his head towards the distance.

"There are some roe deer," he said, using the Ukrainian word *koza*. *Koza* actually means "goat," but it is a shortened version of *kosuli*, the Ukrainian name for roe deer. It is also the meaning of their Latin name *Capreolus capreolus*—"small goat." In fact, the graceful roe deer with their small forked antlers are about the size of goats, which is how I excused my inability to see them despite all of Sasha's pointing.

Everyone else saw them, but they were all holding binoculars while I was holding my notebook and pen—with which I scribbled almost indecipherable notes in the bouncing jeep—and a new digital recorder that I was using for the first time and very much hoped was recording all that my pen was missing.

We got out of the jeep and Sasha kept asking me: "Do you see them?"

But the deer blended in perfectly with the surrounding field. No matter how hard I squinted, I only saw brownish forms that could have been bushes for all that I could tell. I spotted them only when they ran off, their creamy rumps bounding like flags over the brush.

"There they go!" I exclaimed, thrilled to have seen any wild animals at all.

They were, like all living things in the zone, radioactive, though different species display different seasonal fluctuations depending on what they eat. Roe deer's radioactivity levels are highest in the spring and summer, when they tend to feed on grasses instead of their winter diet of buds and shoots. Being physiologically active, buds and new twigs concentrate radionuclides. But grass absorbs them even more because its shallow roots penetrate only the upper layers of soil, where most radionuclides are concentrated.

In a 1992-1993 study of zone game, a kilo of some roe deer's meat contained nearly 300,000 becquerels of cesium-137! This was during the anomalous period of high radioactivity levels that may have been caused by decaying pine needles. Radionuclide concentrations have been dropping since then to an average of 30,000 becquerels in 1997 and 7,400 in 2000, although such levels are still dangerous. In Belarus, the maximum permissible amount of cesium in a kilogram of game meat is 500 becquerels. Ukraine doesn't distinguish between sources of meat and sets a maximum limit of 200 becquerels.

Grazing animals don't only eat radionuclides that have been taken up by plants. They also pick up cesium and strontium directly from the soil that they eat when they pull up plants together with their roots. Indeed, these root layers contain the most radionuclides, including transuranic elements such plutonium and americium that don't get into plants very much through their roots. One reason for this is because they don't mimic any nutritious elements like calcium or potassium. Moreover, plutonium readily and strongly bonds to other atoms to make molecules. But since chemical elements get into plants' roots only in their ionic form, as positively or negatively charged atoms— and not molecules—plutonium's path into plants and into the animals that eat them is usually chemically blocked. Plutonium does, however, get *on* plants when the wind kicks up dust.

With a half-life of 24,110 years, plutonium-239 gets a lot of attention in antinuclear literature because, among other reasons, no imaginable container will remain intact long enough for its safe disposal and storage. But plutonium-240 also has a not-short half-life of 6,564 years, while the half-life of plutonium-241 (about 14 years) means that it is far more radioactive than either of the others.

We were driving through a region where each square meter got sprinkled with about 3,000 becquerels of plutonium-239 and 240. This wasn't too bad, considering that each square meter of the Red Forest got more than 300 times as much, or about 1 million becquerels per square meter.

But you won't find plutonium-241 on the colorful contamination maps, though it is by far the most abundant of the plutonium isotopes that Chernobyl released. The amount of plutonium-241 is from 50 to 100 times as high as the 238, 239, and 240 isotopes combined. But although plutonium-241 can emit an alpha particle, it usually decays by way of a weak beta particle that is hard to detect, making the isotope very difficult to find and map.

Although they are heavy metals and thus toxic, the plutonium isotopes are not, as common myth would have it, the "deadliest substance known to man." They are actually less deadly than some poisons, such as arsenic, that don't provoke such existential dread. Their radioactivity is more dangerous than their chemical toxicity, but even that depends on the type of isotope and where it is located.

Plutonium alpha particles are very energetic. Plutonium-239's alpha is four-and-a-half times as powerful as cesium-137's beta, and its two protons give the particle a double positive charge that rips electrons off neighboring atoms. At the same time, however, the alpha particle is heavy and the nucleus that emits it is surrounded by negatively charged electrons, providing a kind of subatomic friction that drags on the alpha and begins to slow it down almost immediately. External alpha radiation can't penetrate more than a few microns of skin before stopping, picking up some loose electrons and becoming harmless helium. So however counterintuitive it may seem, merely walking around a field of plutonium is actually not very dangerous.

The element does, however, become dangerous if it is inhaled. In the early period after the disaster, when much of the contamination—including plutonium—was on the surface, it could be inhaled easily. Now that nearly all of the radionuclides have sunk more deeply into the soil, the transuranic elements can still occasionally be inhaled, especially in dry and windy weather that kicks up surface dust. But now they get mainly into animals that consume some soil with their food.

The plutonium in nuclear fuel is very insoluble, and the body excretes 99.9999 percent of it if it is eaten. But the fate of plutonium in an animal depends largely on its size. Nano-sized particles, a billionth of a meter in size, are so small that they dissolve into the bloodstream and get metabolized. But since the body has no use for plutonium or anything resembling it, most of the stuff eventually gets excreted, although a tiny fraction of a percent concentrates in the liver and bone. Large pieces, such as specks of dust, are treated like dirt and get excreted or expelled by coughing. The medium-sized particles, a millionth of a meter in size, are the most problematic. There is no reaction in the body to get rid of them, so they tend to stay a long time, shooting alpha particles at surrounding tissue.

In living tissue, which is mostly water, alpha particles usually travel less than the diameter of some cells before stopping. But their energy is so intense that they are like bombs thrown into a small room, destroying everything. Inhaled plutonium is especially dangerous because it lodges in the lungs where it increases the risk of lung cancer in laboratory animals. How much it increases that risk in humans is a matter of some debate, with the nuclear industry citing low numbers and their opponents asserting high ones.

Although all plutonium isotopes can emit alpha radiation, plutonium-241 can also decay by way of a beta particle. When this happens, it produces americium-241, which is much more worrisome than the original plutonium-241. Americium is more soluble than plutonium, which means that it moves more easily with water through the food chain. Its alpha radiation is even more powerful than plutonium's, and it decays to neptunium-237, which also decays by way of an energetic alpha particle and has a half-life of more than 2 million years. The zone's americium-241 will reach its maximum level in 2059, when it will then be more than double the amounts of plutonium-239 and 240.

Unfortunately, americium has been little studied in zone animals because it is difficult to separate from tissue and the small rodents that are the subjects of most Chernobyl animal research don't have much tissue to begin with.

The jeep bumped through a pine forest, strewn with tree trunks and branches that had recently been cleared to make a fire line. The carpet of sand muffled the sound of the jeep, but not enough to stop a group of roe deer from prancing nervously and then bounding off deeper into the forest.

When Yuri Kolesnik spotted some tracks in the sand, we piled out for a closer look.

"*Los,*" he said, using the common Slavic name for "moose." The track was certainly large. The hoofprint's cleaves were about five inches long and almost as wide. Only domestic cattle could have comparably sized tracks, and there were no cattle wandering wild in the zone.

There were, however, some European bison. The largest European mammal was very similar to the North American buffalo, both in appearance and in being brought to the brink of extinction. Bison became extinct in the wild in 1919, but they survived in zoos and have been reintroduced in the Bialowiecza forest, a nature reserve of old primordial woodlands that straddles the border between Poland and Belarus. Some Bialowiecza bison were brought to the Belarusian radiological reserve and released into the wild in 1996. By 2004 there were 37 adults and 3 calves, and the reserve wanted to bring in more since Bialowiecza was getting overpopulated. But all of the bison were on the left bank of the Pripyat River—although they might cross it someday.

Yuri confidently announced that the print belonged to a bull moose, since a female's would be smaller.

He followed the tracks for a bit and then stopped. "Wolves," he said, and we all caught up with him to examine the big canine paw prints. There were a lot of them.

"I've seen a pack of 13 and smaller groups of 2 and 3," said Sasha. "Once I saw a pair just lying on the side of the road. They saw us and then just wandered off."

Farther on, near a pond dammed by beavers, we passed a pine sapling with frayed branches and stripped bark—typical signs of moose, Yuri said—before coming upon a path running parallel to a large field. A section of barbed wire marked the border of the 10-kilometer zone, and somewhere inside Sasha spotted something and ordered the driver, whose name was Boris, to back up to an opening in the barrier.

"You are now illegally entering the 10-kilometer zone," Sasha told me cheerfully as the jeep lurched over a deep ditch to emerge onto a relatively smooth trail.

He was technically correct. According to my program, which named all the places that I would visit on that trip, my point of entry into the "ten" was the checkpoint in Leliv. And we were a good 15 miles from it. But since I did have permission to be in the inner zone, it was unlikely that I'd suffer any consequences even in the highly unlikely event that militia were patrolling the wormwood forests—especially since Sasha, the fire chief, was my escort.

As Boris maneuvered the jeep, Sasha and Yuri looked through their binoculars at some distant specks.

"Red deer," said Yuri.

I squinted and peered through my binoculars but found it impossible to focus in the bumpy jeep. I only saw the deer when it seemed that the entire herd had started to leap across our path.

My recorder preserved my inarticulate reaction: "Super. Wow. My God, they're beautiful!" I had gone on the "safari" expecting to find tracks and spoor. Instead, a herd of red deer was running around my transportation.

The herd crossed our trail as the deer ran from the field into the woods. Twice as big as the roe, red deer are second in size only to the moose. Each antler in an adult stag's crown could be two feet long and weigh more than six pounds, depending on the availability of food. Red deer favorites such as shrubs, bark, tree shoots, and grasses were plentiful in the zone. But the stags had already dropped their antlers for the winter.

Boris revved the engine to catch up with them for a better view.

"There're more," said Sasha, looking through his binoculars and pointing at some distant clumps that hadn't run away with the main group.

I had read several field guides to European mammals before the journey. But if I really wanted to observe wildlife, I clearly had to make a choice between holding my recorder and holding binoculars.

But then I spotted them. They were trying to get away from us, but our noisy jeep probably seemed like a formidable barrier between them and the herd that had already fled into the forest. They trotted off in the opposite direction and soon disappeared from view.

"This is a pantry for boars," said Sasha, when we drove past an over-grown apple orchard in an abandoned village of decaying log cabins and low-slung farm buildings shedding chips of dingy white paint.

Wild boars' seasonal radioactivity level fluctuations differ from those in roe deer. They are relatively low in the autumn, when the boars like to dine on windfall fruits, which don't accumulate radionuclides. Yet while roe eat soil with their food in the warm months, boars do so in winter, when they plow through radioactive forest litter and soil with their snouts in search of roots, small animals, worms, and insects. Boars are the most contaminated of the zone's ungulate species, fol-lowed by roe deer and moose.

One of hottest boars in the Ukrainian zone had 444,200 becquerels of cesium in a kilogram of meat. Belarus beat that with a boar that measured 661,000 becquerels per kilo. But these were both hunted in the early 1990s. As with roe deer, radioactivity levels have been falling since then.

At first, different individual animals, even those from a single herd found eating in the same place, showed widely different levels of con-tamination. In 1992 one boar could measure 40,000 becquerels per kilo, while another had less than 300. Roe deer told a similar story. Because animals move around—and large animals have considerable ranges—a roebuck killed in a highly contaminated area could be rela-tively clean, while another one killed in a clean area could be very con-taminated. Moreover, a boar dining on evening primrose leaves in a heavily contaminated field where most of the radionuclides were em-bedded in fuel particles would be less radioactive than another boar eating evening primrose in a less contaminated field sprinkled with condensed cesium, because the condensed radionuclides get into the plants and the fuel particles don't.

By the turn of the millennium, however, the large animals had be-come more uniformly contaminated and spread all over the zone terri-tory. A boar shot in Leliv in 2002 measured 1,000 becquerels per kilogram, while another one in Ladyzhychi, at the mouth of the Pripyat River five miles away, measured 650. On the left bank of the Pripyat River, a boar shot in a very dirty patch of Belarus that same year had 4,800 becquerels per kilo, while in Otashiv, a clean village on the Ukrainan right bank, a boar shot a year earlier measured 10,000 becquerels. It's possible that the boar swam from some of the contami-nated islands at the river delta.

Yet even 18 years after the disaster, an individual animal can get highly contaminated by eating mushrooms, especially varieties such as porcini with localized mycelia that reach through litter to the mineral layers of soil. One mushroom found in 2002 contained 900,000 becquerels! Of course, that was per kilogram, meaning a boar would have to eat that amount of mushrooms to pick up so many becquerels. But a kilo of mushrooms isn't too much for a boar to eat.

We didn't actually see any boar that day, though we did see plenty of their signs and literally fell into one of their wallows when the jeep got stuck in a deep, muddy ditch filled with water from many past rains that had nowhere to drain because the ground was clay, like the radioactive waste trenches in Burakivka.

We all got out to lighten the load and inspect the damage. The muddy waters reached to well above the wheel hubs, but Boris heroically reached into the glop and switched the front wheels into four-wheel drive before sitting back in the driver's seat and revving the engine. I climbed a sharp, slippery hillock to escape the mud spattering from the jeep's ineffectually spinning wheels.

After lurching back and forth several times, Boris killed the rattling engine and the firemen considered what to do. Though Sasha had his cell phone and the jeep also had a CB radio, everyone preferred to continue the expedition instead of waiting for a rescue winch.

It was highly unlikely that another car would pass by to help us. We spotted only one other car that day, and this was when we emerged onto a paved road for two minutes before descending into the bush again. Altogether, about 100,000 vehicles visit the Ukrainian zone annually, or less than 300 a day. That's about the number of cars that would fit in the parking lot of a midsized American strip mall, spread over an area the size of Rhode Island—though in fact nearly all of the cars are concentrated in the two Chernobyls, the town and the nuclear plant. There are even fewer cars in the Belarusian zone.

"Pile branches into the hole under the front wheel. That should give us some traction," Sasha commanded, as he began collecting the plentiful autumn deadwood. A dead tree stood atop the mound I had climbed, and Yuri clambered up the slope to snap some branches off its bare boughs.

"The boars wallowed and then rubbed themselves on the trunk here," he said pointing to some dark stains on the bark and the deep tracks left by the boar's cloven hoofs.

I also broke branches off the dead tree and threw them into the watery ditch, though the wet clay was very slippery and I shuffled carefully to avoid falling in together with the jeep.

The short November day was on the verge of closing, but none of the firefighters seemed worried. Sasha confidently gave orders, and we piled branches into the ditch while Boris shoveled some clumps out of the ascent to make it less steep. Once we had a large enough pile of branches, Boris climbed back into the driver's seat and started the engine.

It took about five minutes of lurching back and forth, smashing branches into gradually larger footholds for friction, before the jeep finally emerged onto dry ground.

"Things can get worse," Yuri said with a smile after we had all piled back in. "But not often."

Twilight was falling, and we soon spotted four more roe deer that just watched us pass without running away. Then a red deer leaped across the road in front of us to join a larger herd that began running parallel to the jeep. More red deer sprang out of the dusk and ran across our path. Chestnut-colored in the summer, the deer's brown winter pelage blended into the falling darkness, though their creamy rump patches were clearly visible.

"I've never seen this many of them," said Sasha as the jeep curved around a copse of pine trees.

Once plentiful, red deer had largely disappeared from Chernobyl lands in the years before the nuclear disaster. The first immigrants to the zone had probably wandered in from Dymer, a small town about an hour's drive south of Chernobyl, with a large forestry farm that borders the 30-kilometer zone. Even after 18 years however, the zone herds numbered no more than 200 to 300 head. Altogether, we had seen about 25 of them.

Moose, in contrast, are far more plentiful. Although estimating wild animal populations always involves some guesswork, the last count in 2000 estimated about 3,500 moose in the zone. For comparison, northern New York had up to 200 in the year 2000. Alaska had some 150,000, but Alaska is 300 times bigger than the zone.

Yet we were seeing more red deer than moose.

Then Yuri, who had appointed himself moose-searcher, studied a field of bushes and saplings before announcing: "*Los.*"

We all piled out of the jeep for a closer look. But it was getting dark and, if I couldn't spot roe deer in daylight, I could even less see whatever ungulates were the shadowy forms in the distance.

"Those are red deer," Sasha insisted, looking through his binoculars.

Although a bull moose can be twice the size of a red stag, a female moose can be comparable to a red stag—especially if the stags have dropped their antlers for the winter. But wild animals don't stand still to be measured in the wild. If they do, they are usually dead. At a distance and in poor light, they can be hard to identify.

All that I could see were five brownish forms that I could distinguish from the bushes in the field only because they were moving. But Yuri continued studying them through his binoculars.

I started following Sasha back to the jeep when Yuri declared: "They are *los*! I can see their white stockings!"

Sasha peered through his binoculars again and didn't argue with Yuri's identification. After a brief approach towards the forest, the moose evidently decided that we weren't a threat and simply stood there, fading with the light.

They were only shadowy forms to my unaided eye, like wave functions of large deer-like creatures that had not yet collapsed into a specific species. It was as though the firemen's observation had made them moose and I just had to take their word for it. But I had no problem with that. The subatomic world described in this book is observed only indirectly, with dosimeters, scintillators, cyclotrons, and esoteric equations understood only by the shamans of science. And I believe them. Sometimes even very big things like moose can only be seen by those who know how to look.

The mooses' four pale stockings are, apart from size, a way of distinguishing them from red deer, whose legs are brown. Red deer also form matriarchal herds year-round. Stags join them only to rut. Moose, in contrast, are solitary in the summer, and the cows herd together with the males in the winter under the leadership of an alpha female. Young cows give birth to a single calf, but tend to give birth to two as they grow older. Chernobyl cows, regardless of their age, are usually seen with a single calf.

This could be a sign of reduced fertility, which is known to affect other zone animals. For example, wild rodents that spend their lives in the zone have litters of four or so, although laboratory strains of mice and voles exposed to zone radiation produced average litters of seven pups. But unlike moose, whose gestation is 235 days, a vole can have up to seven litters a year in the wild. So, even though Chernobyl voles die at a younger age than their counterparts outside the zone, they also begin reproducing at a younger age, so their population remains stable.

No one knows if similar trends are present with moose because such studies are more difficult and expensive to do with large mammals. All that you need to catch 25 mice is 25 mousetraps, a few ounces of cheese, and a couple of days. Doing the same with, say, moose or boar or deer requires special hunting permits and numerous expeditions that can take months and a good deal of money.

Sampling some animals requires even more than that. For example, because European bison are a protected species, numbering about 2,000 worldwide, the Belarus environmental protection agency must give written permission to shoot one. But although one Chernobyl bison has been lame for three years and scientists at the radiological reserve have been trying to get permission to put it down in order to study its radioactivity levels, the agency continues to say no.

The only comprehensive study of radioactivity in the zone's wild animals, which involved shooting about 50 boar and 50 roe deer from both the Belarusian and the Ukrainian portions of the zone, was funded by the European Union. But when that 1992-1993 study was complete, the Europeans concluded that it was enough.

Belarusian scientists sample zone wildlife regularly, and they even have a charming little hunting lodge in the depths of their wormwood forests. But they hunt no more than 10 animals annually, while Ukraine no longer conducts much scientific hunting at all. A few animals are occasionally killed, but haphazardly. In fact, neither country has conducted any significant large animal hunts since the European study.

PREDATOR AND PREY

The next day I went to Center for the Radioecological Monitoring of the Zone of Alienation, known as the EcoCenter. Once the headquarters for scientists from a variety of institutions, domestic and foreign, studying the impact of radiation on the zone environment, the EcoCenter accumulated a wealth of unique research. But after 2000 the former schoolhouse just a few blocks from Chernobylinterinform became a much quieter place. After cutbacks and layoffs, the EcoCenter had only four people working with wildlife.

Nearly all Chernobyl research had moved to the International Radiological Laboratory in Slavutich, which is one of the make-work projects that the international community (primarily the United States) put money into in the hopes of keeping the former nuclear plant workers and scientists employed, rather than tempted to sell their knowledge and skills to pariah states or terrorists. Unfortunately, no one gave much thought to the fate of the scientists and nuclear experts at the EcoCenter. Few of them got work in Slavutich.

A livestock specialist with a cleft chin and blue eyes, Igor Chizhevsky was one of the remaining EcoCenter experts, and we chatted over tea and cookies in his spacious, well-equipped office.

He handed me a small photo album in which the first picture showed Igor's colleague climbing a tree to a huge pile of sticks on the tree's flat crown.

"It's a white-tailed eagle nest," Igor explained. Large raptors with eight-foot wingspans, white-tailed eagles are very rare in most of Europe. In Ukraine and Belarus, they are listed as endangered. In the zone, though, with its rich supplies of favorite foods such as fish and hares, there are as many as 50 white-tailed eagles. This may not seem like very many, but before Chernobyl, there weren't any. They probably discovered the inviting habitat during their migrations to and from their nesting grounds in Finland.

The photos showed Igor's colleague Serhiy Gashchak from the International Radioecological Institute banding a white-tailed eagle fledgling and attaching a satellite tracking system.

"So, do you know where it is now?" I asked. It was mid-November and most raptors had long since migrated south.

"I don't know if it's the system, or the coordinates, but according to this data . . . ," Igor sighed before continuing, "the last signal we had was from China."

I laughed and so did Igor. "Of course, that's impossible," he said. "The eagles normally migrate to Africa. And the previous signals we had were in Ukraine. So, it must be a mistake."

Whether it was a mistake or not, the eagle's signal disappeared altogether the following year.

I continued leafing through the photo album and came upon a picture of a ruler next to a deep depression in the sand.

"That's a brown bear track," said Igor. "This was last July. The forest rangers spotted it first and they called us. Though there had been sightings of bears and bear signs before, this was the first one that we could document."

No wonder they were all excited. Bears are endangered in Ukraine.

At eight inches in length, it was a big print. With the widest geographic distribution of all bear species, brown bears include the Alaskan Kodiak and American grizzly. The Eurasian brown bear that strolled through the Chernobyl zone was smaller than its North American cousins, but it was still an impressive creature. And it left its print not far from the boar wallow where our jeep got stuck the previous day.

"The print was made the night before we took the picture. We think the bear was a male looking for new territory. It was heading in the direction of Belarus," Igor said.

Bear signs were also reported in Belarus that summer, though none were confirmed, leaving the bear's origins and destination a mystery.

The album also had a series of dusky photos taken with automatic cameras: a badger, three wolves, a beaver, some red deer, moose, boars, and a polecat. A raccoon dog stuck its nose towards the lens, distorting its face like a funhouse mirror.

Originally from East Asia, raccoon dogs look just like what their name suggests—very furry dogs with raccoon-like masks. They are unusual amid canids in being able to climb well, and they are the only canids to hibernate in winter. They swim well, too, and often like to hunt in wetlands, near shores, and in thick reeds. The Soviet Union introduced them in the 1950s as fur animals. Some were deliberately released into the wild, and their descendants became a serious pest in Eastern Europe before their populations stabilized.

The last pictures in the album were of two dead roe deer and a boar, their blood smeared on the floor of Igor's laboratory before he butchered them for tissue samples. The two species are useful to compare because they have different digestive systems. Like cattle and sheep, deer are ruminants that swallow their food essentially unchewed then regurgitate it for additional chewing before reswallowing it. Ruminants' stomachs contain three or four compartments to handle the different stages of digestion. Boars, in contrast, are like people, horses, and domestic pigs. Their stomachs contain only one compartment.

Roe deer and boar dominate scientific studies because they are populous and popular game that can wander out of the zone and pose a danger to people who hunt them for food. There are as many as 3,000 roe deer in the Ukrainian zone and about as many in the Belarusian reserve. And the 7,000 wild boar in both zones represent a 10-fold increase over predisaster years. Their zone numbers might be still higher, but about 600 wolves keep them in check.

Being at the top of the food chain, wolves—like other predators—have very high levels of radioactivity in their muscles.

"Some say that there are too many wolves. But you only hear that from people who want an excuse to hunt them," Igor said. "The population is just right relative to the amount of prey."

Ukraine first allowed limited wolf hunts in the winter of 2003-2004. Belarus, in contrast, has licensed hunts every winter. In the 2003-2004 season, foreigners who paid generously for the opportunity shot nearly 100 wolves from helicopters. Wolves are especially considered a nuisance outside the radiological reserve. A hunter who shoots one wolf gets a free boar-hunting license. The prize for three wolves is a free moose license.

As of the last count in 2000, there were 66 species of mammals living in the Ukrainian zone, including as many as 1,500 beavers, which had virtually disappeared from the area before the disaster, 1,200 foxes, and 300 raccoon dogs. Presumably there were similar numbers on the Belarusian side, though they had not done any animal counts for many years.

Lynx have also appeared in the zone. Decimated in Western Europe about 100 years ago, the long-legged cats with tufted ears are endangered in Ukraine and Belarus. But a 1999 poll of forest rangers—not the most accurate census method, but inexpensive—estimated 15 lynx in the Ukrainian zone, and a similar number are thought to be in Belarus. The zone is one of the few European wild lands large enough to accommodate the lynx's enormous 170-square-mile range. It is also teeming with roe deer, the lynx's favorite food, and hunters, the lynx's worst enemy, are officially banned.

As predators, lynx probably also have high radioactivity levels in their muscle, although no one has actually sampled one to check. But they may have less radioactivity than wolves because roe deer are usually less radioactive than boars, which are wolves' choice prey.

While the animal populations have not yet reached the zone's capacity, the growth in the number of large mammals may be slowing, at least compared to the rapid growth in the 1990s. To know for sure if they are beginning to stabilize, however, the populations would have to be relatively unchanged for five years, and it is not at all clear that either Ukraine or Belarus will provide the funding needed to do the counts in 2005. What is certain is that the numbers are not falling.

Since the health of an animal population is measured by its size rather than the health of all of its individual members (which is practically impossible to measure), then—however counterintuitive it may seem—the huge populations of large Chernobyl mammals are healthy indeed. The same is true of smaller mammals, including rodents.

Not all species are doing well, however. Chronic radiation exposure hits hardest at creatures with long development periods. Those that develop in contaminated soil are especially vulnerable. Maybugs are big beetles that grow from fat white larvae that spend their first three to four years eating roots in the soil. The radiation in zone soil seems to negatively affect their development, though no one knows how or why because finding out would require studies for which neither the affected countries nor the international community has money or interest. All that is known is that there are fewer maybugs in the zone than outside it. Also, the zone's male stag beetles are more asymmetrical than controls, but like the partly albino swallows, crooked stag beetles are not attractive and are less likely to mate.

So, the apocalyptic fiction of a radioactive world inhabited by rodents and insects is not entirely true, at least not in the case of Chernobyl. Indeed, the American cockroach, which conventional wisdom considers a likely survivor of a nuclear holocaust, is actually a wimp among insects when it comes to radiation resistance. To be sure, it is more resistant than humans. But cockroach populations die at levels that other insects don't even notice.

Given the budgetary constraints on scientific hunting in the zone, some radiological information is gleaned from accidental finds such as shed antlers or the remains of animals killed by predators.

"We have a few red deer antlers picked up here and there," Igor explained as I followed him into his laboratory, where he picked up some kind of radiation gadget and placed the sensor—a long wand attached with a wire to a box with dials and gauges—very close to a huge, nine-point antler on a shelf.

"It's clean," he said. "Five or six microroentgens of gamma radiation."

Although exposure is always measured as an hourly rate, most people in Chernobyl just tell you the number of roentgen, without saying "per hour."

The reading was lower than normal background, meaning that the antlers had not accumulated any appreciable cesium-137, whose decay product, barium-137, emits gamma rays.

"But lets check for beta radiation," said Igor and flipped a switch on the detector. "This is higher—170 beta particles per square centimeter per minute."

The maximum permissible level was 20.

A five-point antler that lay next to it on the shelf was slightly dirtier: 400 beta particles per square centimeter.

"The beta radiation is from strontium-90, which metabolically mimics calcium, so it collects in bone, teeth, antlers," said Igor. "Cesium mimics potassium and concentrates in muscle, less so in liver."

"But these were accidental finds. We don't know where the stags that dropped them came from," said Igor. "They could have wandered in from clean places."

Then I asked Igor to check the soles of my shoes, which still had some dried mud on them from the previous day's adventures near the boar wallow.

"Hmmm," he said with the tone of a doctor presented with a worrisome and mysterious symptom. But it meant nothing. The readings were practically zero.

Not so for the collection of animal skulls on the laboratory shelves. A small boar's skull contained 1,300 beta particles per square centimeter. A larger one, from an older animal, measured 2,700.

Then Igor led us to the lab table, piled with some plastic bags. "This is what I'll be working on after you leave—preparing samples for the lab. I can do rough radiation estimates here, but I need the lab for more accurate results."

He unwrapped one bag containing the lower leg of a roe deer. Another contained the remains of a headless hoopoe found in a relatively clean spot near the mouth of the Pripyat River. Though its colors are subdued and unobtrusive on the ground, the hoopoe is transformed in flight, revealing a dazzling wing pattern of black and white. It also has a fan-shaped crest tipped with black that looks like an American Indian chief's elaborate headdress when opened. Hoopoes are unlike any other

European bird. The first hoopoe I spotted during a picnic outside Kiev led me to take up birding as a casual hobby.

Of course, I had to ask Igor the obvious question, the question that I get asked whenever I tell people about my travels to Chernobyl and one I've repeated to nearly every scientist I've interviewed about zone wildlife.

"So, have you seen any mutants?"

But the answers are invariably the same.

"No," said Igor.

"C'mon," I exclaimed. "Everyone knows that radiation causes mutations. How can it be that there are none in Chernobyl?"

"Because with wild animals, mutants die. If they actually are born, we never see them because scavengers eat them before we get a chance," Igor responded. "Only the individuals that can adjust to the conditions here survive."

I recalled the eight-legged colt, dubbed "Gorbachev's colt" after a Ukrainian scientist brought a life-size photo of it to Moscow in 1988 to show Mikhail Gorbachev what Chernobyl was doing to the country's animals. Actually, no one knew for certain if the disaster caused the deformities. But if it did, the colt survived long enough to be politically useful only because it was born on a collective farm and not in the wild.

Not all mutations cause gross deformities. Certain Chernobyl wasps, for example, display more variety in the patterns on their bodies than wasps outside the zone. Yet when it comes to mammals, even genetic changes, with effects invisible to the naked eye, have been minimal. One 1996 study of Chernobyl rodents reported high rates of genetic mutation in two species of voles, but the increased rates turned out to be mistakes and the authors retracted their conclusion a year later.

Other studies have found a slightly higher number of mutations in the mitochondrial DNA of Chernobyl bank voles. Mitochondria are the tiny subcellular structures that generate energy. They have their own snippet of DNA, probably because they were once free-living bacteria that set up shop in more advanced cells hundreds of millions, perhaps even billions, of years ago. But even their mutation rate is not statistically significant. Thus far, no one knows why there haven't been more genetic changes.

Igor showed me an entire freezer drawer packed with spadefoot toads he collected in the Red Forest during the summer. Frozen into round clumps, the toads were among the first zone amphibians to be studied. In general, amphibians are environmental bellwethers because their unshelled eggs and permeable skin make them hypersensitive to environmental perturbations. But amazingly enough, no deformed frogs have been found in the Red Forest.

You are more likely to encounter a deformed toad or frog in the United States, where there has been a shocking increase in frog and toad malformations, especially missing or extra legs. One cause may be a natural parasite whose populations bloom when runoff from fertilizer, cattle manure, and other contaminants gets into ponds where the frogs develop. The parasites lodge on tadpoles, forming cysts that disrupt the growth of their limbs. But because the only cattle in the zone are experimental and agriculture is banned, Chernobyl frogs don't have that problem. Although excessive ultraviolet radiation can also cause deformities in frogs, for now it seems that runoff is worse than radiation—at the least the type of radiation found in the zone.

"The population and diversity of small creatures in the Red Forest are the same as in comparable places that are less radioactive," said Igor. "If there are differences, they are based on factors other than radiation."

Indeed, although the charismatic megafauna I saw on my safari were most exciting to watch, the vast majority of radiology research on Chernobyl wildlife has focused on animals with far less star appeal: rodents. Not only are they plentiful and easily caught, the little creatures play a large role in the movement of radionuclides through the food chain.

LINKS IN THE FOOD CHAIN

Although the doses to wild animals in the early months after the disaster were largely from external radiation, internal radiation exposure has grown in importance over the years as radionuclides washed off surfaces and into the soil. Some remained in the ground in the form of fuel particles; some decayed away. Sticky clay particles, bacteria, and other elements in the soil adsorbed others.

Part of what remained, especially the soluble and condensed cesium and strontium, followed the flow of energy and the nutrients that they mimic in the ecosystem. Autotrophs such as higher plants, various groups of algae, and certain protists and bacteria that harness the sun's energy to photosynthesize their own food, pick up radioactive strontium and cesium chemically disguised as calcium and potassium. Heterotrophs, which can't manufacture their own food and must eat autotrophs for nourishment, pick up their radionuclides as well.

Herbivores get nutrition by eating living plants. Carnivores and parasites get it by feeding on other animals. The animals' droppings are food for coprophages, or dung eaters, while their bodies eventually provide food for the saprophages—various insects, worms, microorganisms, and fungi that dine on the dead.

Dust to dust, it all ends up back in the food chain.

Actually, since most organisms eat more than one kind of food and are prey for more than one kind of predator, it is more accurate to speak of a "food web" rather than a "food chain." One field vole can eat some grass and in turn be eaten by an owl that eventually dies of old age. Another vole can gnaw on bark and become dinner for a fox, which in turn becomes breakfast for a wolf.

Since the concentration of nutrients grows with each link of the food chain, so do concentrations of some radionuclides such as cesium that imitate them. It has long been known that large predators generally accumulate the most toxins and contaminants, including radioactivity, in their food chains. But other food chains also concentrate radionuclides with each link. In fact, insects that eat dung and those that eat dead animals accumulate radionuclides at rates comparable to those of large carnivores.

Of all the creatures great and small in the zone, the large animals—regardless of their diet—can accumulate the highest *amounts* of radionuclides simply because their size provides more storage capacity. But rodents actually receive the highest *doses*. Indeed, rodents, especially mice and voles, have higher radiation doses than large animals precisely because they *are* small and, like maybug larvae, spend much of their time in radionuclide-laced soil.

A typical vole eating grass on a moderately contaminated pink patch of the radiation maps, where cesium contamination levels are 50 to 100 curies per square kilometer, accumulates up to 44,000 becquerels of cesium in its eight- or nine-month lifetime.

Now, cesium-137 decays to gamma-emitting barium-137 by emitting a high-energy beta particle that can travel about half an inch in biological tissue. In the cesium-packed muscles of a moose, that half-inch is more likely than not to remain within the muscle tissue. In the tiny body of a vole, however, half an inch in any direction is likely to lead to a vital organ.

A similar story holds true for external radiation. Whereas gamma rays penetrate all living things with equal opportunity and heavy alpha particles barely penetrate anything at all, the damage that external beta radiation can do differs among beta particles—because of their different energies—and the species of animal they hit. A boar wallowing in the mud of a pink patch of the zone can be exposed to beta radiation from all sides. But even cesium-137's most energetic beta particle won't penetrate much deeper than its fur and maybe a few millimeters of muscle, depending on which part of the boar's body it hits. Because boars don't wallow all the time, however, their external beta exposure is often more localized, say, from the soil to their hoofs or from the bark of a tree they rub against.

Voles, in contrast, often nest in underground tunnels and spend much of their outdoor life in direct contact with the soil surface, putting them in direct and constant contact with environmental radionuclides. Because of the animals' small size, beta radiation from external sources is much more likely to hit an internal organ.

All of this damage may explain why Chernobyl rodents have shorter life spans and smaller litters than their counterparts outside the zone. Rodents from the 10-kilometer zone have more pathogen colonies on their skin than controls, indicating that their immune systems are depressed. But this does not affect their numbers or their disproportionately large impact on radionuclide recycling in the ecosystem.

In fact, given their numbers, reproductive potential, and short life spans, the average population of rodents in one moderately contaminated hectare of the zone processes 3.5 million becquerels of cesium annually. This amounts to only a tiny fraction of a curie, but if you multiply that fraction by the Belarus and Ukrainian zones' total area of about half a million hectares—and ignore the patchiness of the radioactivity levels, which will affect how much cesium the rodents take up—it turns out that rodents process about 50 curies annually. This is a large and significant amount because the cesium becomes biologically active and readily taken back into the food chain when it leaves the animals' bodies, in either the excrement from living rodents or the decomposed tissue from dead ones.

Unlike mice, large animals have large ranges, a fact that has different food chain implications. Game animals wander outside the zone's borders where they can be shot by hunters or leave droppings laced with radioactive cesium. Boars are particularly peripatetic. Igor's friends in his hometown of Ivankiv, a 20-minute drive from the zone's southern border, once shot a boar that measured 6,000 becquerels of cesium in a kilogram of meat.

"I told them not to eat it," he said.

Igor himself once shot a roe deer measuring 1,000 bequerels in the hunting grounds outside Ivankiv. "I ate that one," he said with a mischievous smile.

"You're kidding, right?" exclaimed Chernobylinterinform's Rimma Kyselytsia, who had come by to take me to my next appointment. "The maximum allowable levels of cesium-137 for meat are 200 becquerels per kilo."

The maximum permissible amount of strontium in a kilogram of meat is 20 becquerels. In fish it is 150 becquerels of cesium per kilogram and 35 for strontium, while in fruit it is 70 becquerels of cesium and 10 of strontium. The maximum amount of cesium allowed in a kilogram of mushrooms or berries is 500 becquerels of cesium and 50 of strontium.

"That's in Ukraine," said Igor. "But Ukraine practices overkill when it comes to radioactivity levels in food. The international standard for permissible radioactivity levels is 1,000 becquerels and that's good enough for me."

The United Nations set the 1,000-becquerel standard for cross-border trade in food, but other countries also set lower thresholds— though not as low as Ukraine's. In Japan it's 350 becquerels per kilogram. In Europe it's 600. In Belarus the maximum cesium level allowable in game meat is 500 becquerels.

Soaking meat in brine for an hour can remove nearly half of the radionuclides, while soaking for a day gets rid of more than 80 percent, though it also leaches out vitamins and nutrients.

Rimma looked at Igor skeptically.

"Hey, I've never measured outside the norm," he said defensively.

All zone workers annually undergo mandatory physicals, including measurements of their internal radioactivity.

"The last time I had 70 becquerels," he said.

"Well, I don't eat game and I had less than zero," said Rimma. "It was within the margin of error."

"My highest was 700 becquerels," said Igor as if it was a game of one-upsmanship. "We had this tradition when we went boar hunting under the European research program in '92-'93. At the end of every season, we ate a boar piglet. I got measured very soon after one of those meals."

Cesium, like the potassium it imitates, doesn't stay in the body but turns over constantly, reaching equilibrium after about 100 days. This means that for any new cesium or potassium entering the body the same amount is excreted. So, if Igor ate only clean food after consuming the piglet, potassium would gradually replace the radioactive cesium, cleansing him in about three months. The same is true of livestock. Meat from cattle pastured on radioactive grass will be clean if they eat uncontaminated fodder for three months before slaughter. Migrating birds lose their cesium in their wintering grounds.

Strontium, in contrast, is like calcium and keeps building up in the body for many years. So as a rule, the older the individual (of whatever species), the higher its strontium count.

This could explain why differences in the amounts of strontium in individual Chernobyl rodents are lowest in the winter. In summer the scientists collecting samples catch both older individuals, who have accumulated a lot of strontium in the course of their lifetime, and young ones, who haven't. So the differences between them can be large. In winter, however, the older rodents tend to die and the live rodents caught in the scientists' traps are all younger and, thus, closer to the lower end of the strontium scale.

After living in Kiev for 12 years and buying food at unregulated farmers markets (or right from farmers' doorsteps), I wondered if I had internalized any radioactivity on my own end of the food chain. So Rimma took me to a surprisingly pleasant room in the Chornobyl polyclinic where Vasilina Puchkova, a chatty and cheerful engineer, told me to sit in a red vinyl chair attached to a scintillator that measured the radioactivity emanating from my chest.

I had to sit still for about two minutes with my back pressed against the chair. But it was an entertaining wait. Puchkova had a mini-museum of Polissian arts and crafts that she had collected from abandoned villages over the years: embroidered linen towels and hemp blouses, silvery samovars, a spinning wheel, and dozens of wooden implements whose purpose I could only guess.

After three minutes, a picture of my internal radioactivity appeared on Puchkova's computer and we gathered around the monitor, which displayed a graph showing the energy of the radiation on the x axis, and its activity on the y axis. I experienced a moment of trepidation when I saw the mysterious peaks and troughs, but then Puchkova assured me: "Not to worry" (Figure 3).

"This is where your cesium peak would be, if you had any," she said, pointing to a flat part of my graph. "But you don't. Well, maybe you do, but the amount is too small for this instrument to measure. It only gives a rough estimate." Then she pointed to the sole peak on the graph. "This is potassium-40. Everyone has this and it's not counted in the results."

A typical person weighing about 150 pounds contains about 17 milligrams of potassium-40. Like the natural uranium isotopes, it is a primordial radionuclide—a chemical echo of the cosmos's creation. It is weakly radioactive with a half-life of more than a billion years. Other natural radionuclides in our bodies are not primordial but, like carbon-14, are created by cosmic rays zapping elements in the Earth's upper atmosphere.

The numerical results were in the form of a becquerel count from the excess Chernobyl cesium, but my count was *minus* 7 or 8 because the program made its estimate on the basis of the average Ukrainian's

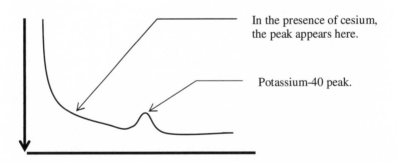

In the presence of cesium, the peak appears here.

Potassium-40 peak.

FIGURE 3 Mary's internal radioactivity chart.

body weight for a given height and age. Since I am thinner than the average forty-something, five-foot-eight-inch Ukrainian woman, I ended up with a negative reading. It was a pleasant surprise. I might have had more cesium from living and eating in Kiev in the early 1990s, but potassium, radioactive and otherwise, had replaced it.

While we were there, Rimma got herself tested, with a result of 100 becquerels rather than the zero she had been expecting.

"Where'd that come from?" she wondered.

Another woman who came in while we chatted with Puchkova wasn't so lucky. Her graph showed a cesium-137 peak that measured 2,822 becquerels.

"That's not bad at all," Puchkova said when the woman left. "It's only when the readings exceed 37,000 that we have to report it to the zone administration, and even then, it doesn't mean that the person will be banned from working here."

"See, Rimma, 100 becquerels isn't so bad," I said, filled with confidence over my negative reading, though rather concerned about the high internal radioactivity readings that were tolerated in zone workers.

"It's nothing," Puchkova told Rimma. "It's within the scintillator's margin of error."

"I know. I'm not worried," Rimma responded. "I'm just wondering where I got it. I haven't eaten any mushrooms." After thinking a few moments, she said: "I know! The last time I visited the *samosels* they treated me with some honey. Bees can carry a lot of radionuclides."

Covered in tiny branched hairs that make them like flying balls of Velcro, honey bees are like nature's dust mops. They pick up bits of everything while foraging and bring it all back to the hive, making them excellent, cheap, and fast environmental sentinels. Bees are so efficient at environmental monitoring that radioactive cesium will be detectable in their hives even when its levels in the environment are negligible. The flip side, of course, is that their honey concentrates radionuclides.

"How about strontium? Do you measure that, too?" I asked Puchkova. Even if my body had already expelled some old radioactive cesium, any strontium I might have picked up over the years would still be lurking somewhere in my bones or my teeth. Germany recorded a 10-fold increase in strontium-90 in baby teeth after Chernobyl. Of course, I was an adult and because my teeth were already formed, they were less likely to have absorbed any strontium-90 than a child whose teeth were still growing.

Puchkova shook her head. "No one has figured out how to accurately measure strontium in living tissue yet. There have been some experiments, but you have to lay there for something like eight hours and the results are still unreliable."

Strontium-90 was difficult to measure in the body for the same reason it is difficult to measure in the field. Its beta radiation can't be distinguished from beta radiation from other isotopes without taking physical samples and doing a lot of cooking in the lab. But I wasn't about to let anyone take samples of me.

And so I left, reassured about my cesium levels and totally in the dark about my strontium. It was unsettling. Like a roebuck browsing on tasty buds in the wilds of the zone, I was clueless about the radionuclides that may or may not be in my body. Except the roebuck doesn't know what it doesn't know. I knew. And that made us very different kinds of links in the food chain.

5

Back to the Wild

Man, alone, violates the established order and, by cultivation,
upsets the equilibrium.

—Vladimir Vernadsky

Chernobyl is far from the only place on the planet that human beings have rendered uninhabitable. War can leave formerly civilized lands open for wildlife. Tigers rebounded in Southeast Asia during the Vietnam War because combat drove out the farmers that killed them to protect their families and livestock. The kouprey, a wild cow and one of the rarest animals on the planet, may have also benefited. Thought to be extinct, kouprey were spotted in the Vietnamese highlands in 1988.

Preparation for war also excludes people from swaths of territory surrounding military bases and weapons production facilities. The Savannah River Site, which produced plutonium and tritium for nuclear weapons, carved 350 square miles out of South Carolina that have been virtually undisturbed for decades. In 2001, federal lands around the Rocky Flats nuclear facility in Colorado were officially declared a national wildlife refuge.

Buffer zones between warring tribes or countries can also create no-man's-lands. Borderlands between Native American tribes in the American West may have preserved big game such as buffalo and elk from the extinction that other North American megafauna suffered after the arrival of humans 13,000 years ago. In the last century, the demilitarized zone between North and South Korea became a patch of paradise. Like the flaming sword that God installed east of Eden to prevent man from reentering after the expulsion of Adam and Eve, land mines and machine guns kept people out of the DMZ, making it welcoming for wildlife.

Science fiction writer Bruce Sterling coined the term "involuntary parks" to describe places that have been reclaimed by nature and reverted to savagery because of war, pollution, or other disasters. Involuntary parks do not, as Sterling puts it, represent untouched nature, but "vengeful nature"—natural processes reasserting themselves in areas of political and technological collapse.

Few terms have struck me as more appropriate to describe the Zone of Alienation around Chernobyl. Even the name, bestowed by bureaucracy, conjures the zone's rejection, its indifference to humans.

While civilization with one hand destroys habitats and leads to the extinction of species, it can also do the opposite with the other. University of Arizona paleoecologist Paul Martin, for example, advocates the creation of nature preserves that would re-create the pre-Indian, Ice Age world with African or Asian elephants standing in for the extinct North American elephant species. Bruce Sterling goes further and, his tongue only partly in cheek, suggests that DNA from frozen mammoths in Siberia be used to resurrect mammoths—if and when that becomes technically feasible.

Sterling maintains that civilization has done so much damage to the environment, including climate change, that we can't rely on nature to reassert itself without some ambitious interventions. Extinct species such as mammoth will not resurrect no matter how many voluntary or involuntary parks are created. While resurrecting extinct species, even if it becomes possible, poses problems of its own—woolly mammoth would probably find it a tad warm if global warming proceeds apace—Chernobyl offers surprising examples of species that were brought to the brink of extinction and are now thriving in the wild.

FAUNA

It was a snowy, windy, and freezing December day at the close of 2003 when I once again found myself off-roading in the Chornobyl fire department's UAZ jeep. It was actually the last place I wanted to be at that particular time. But two scientists that I had been eager to meet for months had come to the zone for a week-long visit. Since tight funding allowed them to visit only once a year, I couldn't pass up the opportunity.

Sasha Berovsky, the Chernobyl fire chief, was in front with the driver, Mykola, while I sat squeezed between Tatiana Zharkikh and Natalia Yasynetska in the back. The two scientists were from Askania Nova, a nature preserve in southern Ukraine, which is one of the few places one can see what the natural steppe looked like before cultivation and industrialization destroyed it. It is also the world's largest breeding facility for endangered Przewalski's horses, a stocky species of wild horse that survived only in zoos until recently.

"We breed some Ukrainian Riding Horses, too," said Tatiana, a thirtyish biologist who has worked at Askania Nova since 1992. "It's a unique breed. Have you heard of it?"

Actually I had. The Ukrainian Riding Horse was one of a number of breeds developed in the Soviet Union whose existence was virtually unknown in the West until the USSR collapsed. But I wasn't sure if I knew what exactly was so unique about the breed, so I asked Tatiana.

"It's the combination of their good athletic qualities with their endurance," Tatiana responded.

"And their ability to live in poor conditions," added Natalia, a petite geneticist who has been working with rare animals, including the Przewalskis, at Askania Nova since 1975.

In Soviet times, the poor conditions were due largely to the shortage-ridden economy. In Ukraine the problem was money—especially in the early years of independence when the state-run economy crumbled and a market had not yet formed to replace it. Thousands of horses were slaughtered because there was simply no way to feed them. The best were exported for pennies to Europe.

The scientists had come to the zone to check on some Przewalski's horses that were released into the wild in a controversial experiment several years earlier. I had met with some of the program's opponents in Kiev and was curious about how Tatiana and Natalia would react to their critics.

"One former zone biologist told me that the real reason it was done was because Askania had too many of the horses for its facilities," I said. "Since the horses are endangered, they can't be killed, so you brought them here instead."

"That's not true," Natalia insisted. "Ukraine has no more and no less captive Przewalski's horses than its facilities—in Askania Nova and in parks—can accommodate. All over the world, the problem with endangered and rare species bred in captivity is that there is no place to let them go free."

This was logical. Lack of space and forage shortages were also problems for the bison in the Bialowieza reserve, which is why some animals were brought to the Belarusian zone.

One of the main reasons for species extinction is habitat extinction. The European bison used to roam the continent in great herds in Roman times, but as their primordial forest habitat shrank, so did their numbers until the species was pushed into its last refuge in the Bialowieza forest. This is why the bison are being reintroduced in the same place that was their last home. Unlike the American buffalo, which were plains creatures, the European bison is a woodlands species.

The habitat of Przewalski's horses has also been shrinking steadily since the last Ice Age, when northern parts of Europe were covered by ice sheets and good portions of the rest were arid steppe. Artists of that ancient era painted caves in France, Italy, and Spain with more than 600 depictions of equines resembling Przewalski's horses. But when the ice retreated, the steppes in much of Europe became forested. The exception was a wide strip of territory that extended from the Hungarian plain, through southern Ukraine and eastward into Eurasia, where wild horses were abundant into Neolithic times. Indeed, horses were first domesticated on the steppes between the Dnieper and Ural Rivers sometimes around 3500 B.C.E., though it is not clear if the horses were Przewalski's or Tarpans. More similar in appearance to domestic horses than Przewalski's, Tarpans were wild horses that lived in Ukraine and Eastern Europe until the last one died near Askania Nova in 1897.

Little was known in the West about the Przewalski's before their "discovery" in Mongolia in 1878 by Colonel Nikolai Przewalski, a Russian explorer of Polish origin in the service of the czar. Of course, the Mongolians knew all about the horses, which they called *takhi* and hunted for food. But the discovery of a previously unknown species of wild horse, named in the explorer's honor, sparked the interest of European zoos and animal collectors. First among them was Baron Friederich E. von Falz-Fein, a German-born landowner with an estate called Askania Nova in the steppes of what was then southern Russia and is now southern Ukraine. He sponsored the first expedition to capture the horses.

Because adult horses proved too fast to capture, the expeditions focused on foals, but the first captives died when fed sheep's milk. Later expeditions became more successful once the hunters realized that domestic mares could nurse the wild foals, and the first Przewalski's foals were delivered to Askania Nova in 1899. Altogether, four expeditions caught 53 foals that survived and were sold to Askania Nova as well as other zoos and collectors. But although Askania Nova—which the Communists nationalized and turned into a state nature reserve—was successful in breeding the horses, producing 37 foals over 40 years, the horses generally did not breed well.

Many Przewalski's were also killed during fighting in World War II. When it was over, only 31 horses were left in captivity. And those survivors became extremely important in the 1960s, when severe weather, expanding livestock pastures, and an increase in horse hunting led to the Przewalski's horses' extinction in the wild. In 1960 there were only 59 Przewalski's horses in the world.

With careful husbandry, however, the Przewalskis became a captive breeding success story. By the mid-1980s, there were more than 600 horses, enough to consider experimentally releasing some into the wild. As of 1999, there were more than 1,500. With 89 head in 2003, Askania Nova had the world's largest collection. But although there were relatively plenty of horses worldwide, there were few suitable wild habitats for them.

"At Askania, we developed a release program if appropriate places could be found," Natalia said. "After all, Ukraine, including Polissia, had been home to Tarpans in the past."

"We thought to do it in the Crimean Mountains, where there are good pastures," added Tatiana. "But the Crimean Tatars have been taking over the land there and using the pastures for domestic livestock."

Deported en masse to Central Asia by Stalin in 1944, Crimean Tatars have been returning to their ancestral lands in what is now Ukraine since the waning days of the Soviet Union. But since their homes and lands were largely settled by ethnic Russians that Stalin sent to Crimea to replace the deportees, many Tatars have been taking whatever lands are left. Unfortunately, the Crimean Tatars suffer one of many Soviet-era injustices that can't be easily remedied.

"Then, in Chernobyl, they came up with the idea of introducing different animals to the zone," Natalia continued.

The program, called "Fauna," was the brainstorm of some Chernobyl forestry officials concerned about "excess vegetation" creating fire hazards on the zone's fallow lands. They suggested bringing in Przewalski's horses, European bison, even beef cattle, that would eat the grass, trample the ground, and instinctively and cheaply perform other environmental management duties. The idea had a certain logic but for the fact that the zone had already become populated with plenty of wild animals. The program met with fierce opposition from many zone specialists in Ukraine. In Belarus, in contrast, there was little dissent over the bison program—perhaps because dissent in general was less tolerated in that country.

The Askania Nova scientists were cautious at first.

"But since we had been looking for an appropriate place to release small herds into the wild, we decided it was a good idea," said Natalia. "After all, if this is a place where the horses can be free, why not try it as an experiment? There are places in the zone where the radioactivity is not very high."

"Look!" Sasha interrupted when a lone red deer ran across our path.

"Make a note, Natalia," said Tatiana, who then explained to me that they were also assessing the number and status of other ungulates in the zone.

Given the fact that all of these ungulates are radioactive, I asked if they had done any studies of radioactivity in the horses.

"We sent samples from a mare that died of natural causes to the Institute of Agricultural Radiology outside of Kiev, and all that they told us was that the radioactivity levels were within established norms," said Tatiana.

That was surprising. Horses—like roe deer—eat grass, which is highly radioactive.

But at least the Askania Nova scientists got to test a specimen. As of 2004, Belarus scientists couldn't directly measure the radionuclides in the zone's bison and could only estimate it indirectly, by measuring radioactivity levels in their food and manure.

When I asked if the small world of Przewalski's horses professionals criticized the idea of introducing the horses into a radioactive zone, Tatiana sighed before answering: "You can't even imagine. But Askania

Nova isn't pristine either. There are few parts of the Earth that aren't damaged."

"Then we saw the tremendous number of hoofed animals already living in the zone—red deer, elk, boars," Natalia added, as the jeep bumped and jolted across the brush. "There was even a herd of feral domestic horses that were left behind after the evacuation, but they were taken away after the Przewalski's program started to prevent cross-breeding."

Though it is sometimes said that Przewalskis were the ancestors of domestic horses, genetic studies suggest that the two species diverged before horse domestication.

Przewalski's horses have 66 chromosomes and domestic horses have 64, but the genetic difference between the two is evidently smaller than that between, say, domestic horses and donkeys. When horses and donkeys cross-breed, their offspring—mules—are sterile (with exceedingly rare exceptions). But Przewalskis bred with domestic horses produce fertile young.

"If all those animals were doing well, there was no reason why our horses wouldn't either. After all, in Askania Nova we have to strictly limit the horses' breeding—even that of the best specimens. There's simply no space. But here they can be free and do what nature intended—reproduce. Then, after we brought the horses in and saw that the foals born in the zone survived the winter very well, we saw that the program was a success and we brought in another group of horses."

Altogether, 31 horses—28 from Askania Nova plus 3 from a small zoo on a Kharkiv stud farm—were brought to the zone in 1998 and 1999. Ten of them died, mostly in transport. Although domestic horses can also panic in transport, semiwild Przewalski's horses can literally die from the stress of being cut off from their herd, locked into a tight and dark crate, and driven in a truck for the 15 or so hours it takes to get from Askania to the zone. Young horses are especially vulnerable.

The 21 horses that survived were initially kept in a 100-hectare semireserve enclosed by a six-foot fence to adjust to the new climate. Polissia is colder and wetter than the steppe to which they had grown accustomed, and the horses needed time to figure out what was edible and what wasn't, while getting supplemental feed rations from the Fauna program's rangers. They ranged freely for months, using a former poultry farm's barns for shelter and a small natural pond for water that they shared in the summer with wild ducks.

Not everything went according to plan. One December day an Askania Nova stallion named Volny, or "Free," broke the barriers, herded some unattached mares, and took off. Since it was too soon to release the horses into the wild, the Fauna rangers herded them back into the enclosure and fixed the fence. But Volny, who fully lived up to his name, broke the fence again and freed his mares for good.

Another group of mares had formed a family group with Pioneer, a seven-year-old stallion from the Kharkiv stud farm. But another Askania Nova stallion named Vypad, whose name means "Attack," also lived up to his name by beating Pioneer mercilessly and then stealing his mares. As if that wasn't enough, Vypad broke the enclosures several days later to beat up Pioneer again, for no reason except he evidently didn't like Przewalski's stallions from Kharkiv. Vypad didn't attack any Askania Nova stallions.

After that, poor Pioneer was terrified of Przewalski's stallions and hid if they came anywhere near the enclosure. He ended up covering domestic mares, but the resulting hybrid foals were eventually sent to slaughterhouses.

"But aren't you supposed to avoid hybridization in these introduction programs?" I asked the scientists after they told me the story. In fact, all the Przewalski's alive today have a drop of domestic blood from a mare that served as a foster mother to foals transported from Mongolia. One aim of Przewalski's breeding programs is to dilute that drop as much as possible.

Natalia hesitated and answered carefully: "Any program has those who plan and those who execute. And one of the people in the working group for some reason thought that he was the first person to try hybridization and was determined to do it, though we insisted that it was a bad idea."

In the end, none of the Kharkiv stallions ended up with mares or in the wild. Kept in small groups in the stud farm's zoo, they were practically domesticated and learned few of the social skills needed to do what stallions must in the wild. Askania Nova, in contrast, keeps the bachelor stallions in large herds of up to 30 head where they learn to fight—for food, for their place in the hierarchy, and eventually, if they are lucky, for mares. By the time they were brought to Chernobyl, the stallions were experienced adults.

They took to the wild with relish. The herds rarely returned to the place of their semifree captivity even though it was within their range. Instead they seemed to be expanding their territory.

By December 2003, the wild population had grown to 65 head— two-thirds of them born in the zone—and divided into three groups. Two groups are family herds, called harems, of mares and their young offspring, usually (but not always) headed by a stallion. Volny led a smaller harem of five mares and their youngsters. But Vypad led the largest herd of eight breeding mares.

The remaining horses formed what is called a bachelor herd, which—as suggested by the name—contains stallions that have not yet managed to collect any mares. Like a gang of guys on the prowl, the bachelor herds are less stable than the harems, as stallions occasionally run off in search of mares and then return again to hang out with the boys. When they were first released, some bachelors were able to kidnap domestic mares kept at various zone services until the domestic horses were rounded up and removed. So, hunting farther and farther afield for mates, stallions have been sighted outside the zone, and their spoor has even been spotted in the Belarus part of the zone, which they probably reached by crossing the frozen Pripyat in winter.

The herds today are essentially on their own. The Fauna program and its funding were shut down in 2000, so the only people keeping track of the horses are the two Askania Nova scientists, whose tiny budget lets them visit the zone only once a year, and Sasha, the fireman, who keeps a regular watch and squires the scientists when they visit.

"He's the top stallion!" said Tatiana. "Without him, we wouldn't be able to do anything here. And he does it on pure enthusiasm!"

Blushing from the praise, Sasha laughed and then modestly changed the subject. "The horses have everything they need here. Not just food, but space—there's no other place in Ukraine where they can have so much territory."

"In Askania, there's only enough pasture for about 90 head," he added. "And here, the pasturage is. . . ."

"Everywhere," I said, wiping at the window to gaze at the fallow fields carpeted with thick piles of yellowed grasses, lightly frosted with snow. With freezing temperatures outside and five people inside, the jeep's windows kept fogging up. It also smelled of exhaust and gasoline, from a canister behind the back seat.

"How about predators?" I asked. "Wolves?"

"They're not afraid of wolves," said Sasha. "Wolves are afraid of them."

"Actually, the wolf problem or lack thereof depends largely on the herd stallions," Tatiana said. "Their main job is to protect the herd, and especially vulnerable foals, from predators. Vypad and Volny haven't lost a single foal."

"But if they have no natural enemies in the zone, won't they overpopulate?" I asked, recalling disastrous species introductions such as rabbits in Australia.

"No," said Tatiana. "Because even though there's enough pasturage for thousands of horses, their populations will be limited by the fact that Przewalski's herds don't like to be close neighbors with other herds. By our calculations, each Przewalski's horse in a herd needs 20 hectares of territory a year. That's far more pasture than the horse is able to eat."

The remaining pasture was what each horse needed for its personal space. Added together, it was what each herd needed for its collective space. And each space had definite borders. The zone's two family herds were separated by a six-mile border of no-horse-land that neither entered.

Once the Askania Nova scientists saw that the horses were doing well in the wild, they had hoped to bring in more horses to increase the genetic diversity.

"The international practice with introductions is to release the horses in family groups of a stallion and five mares," said Tatiana as we drove around a small lake surrounded by picnic tables placed there by forest rangers. "But we now think its better to limit each stallion to two mares, while increasing the number of small herds so that more stallions can take part in reproduction."

Unfortunately, the Fauna program died before they could test their ideas.

HOME ON THE RADIOACTIVE RANGE

We were driving through Volny's home range just west of Chernobyl. At 170 square kilometers, it was, according to Tatiana, the largest home range of any wild Przewalski's herd in the world. The size is even more impressive when you have to trundle through it for many, many hours in a bouncing jeep to find the horses. We had been traveling for several hours, with no sign of them. The occasional mounds of manure we spotted were cold and dusted with snow.

"In the early years after we released the horses, seeing manure meant you were about to see the horses," said Natalia. "Now, their ranges have gotten so large that it doesn't mean anything."

Volny's herd's range was about half the size of Vypad's. It also covered the most radioactive parts of the 10-kilometer zone. After kidnapping his mares in December 1999, Volny led his new harem to the buried village of Kopachi. Aside from being one of the most radioactive places in the 10-kilometer zone, Kopachi is also one of the busiest places outside the nuclear plant and the town of Chornobyl because the road between the two runs right through it. So, the following spring the herd moved to the area around the Red Forest and the highly contaminated western arrow of plutonium-embedded fuel particles that exploded from the reactor on April 26, 1986.

The colorful radiation maps depicted the plutonium contamination in the shape of a swallowtail butterfly—a rare species that is now common in the zone—with gradually fading wings extended from either side of the power plant. The deepest hues on the western wing, reflecting contamination levels between 400,000 and 1,000,000 becquerels per square kilometer, were right in Volny's range.

Like plutonium, the uranium-235 and 238 in the fuel can also emit alpha particles. But because plutonium has a shorter half-life than either uranium isotope and because its alpha particle is much more powerful, plutonium-239 is hundreds of thousands of times more harmful per gram than uranium-238 and tens of thousands of times more harmful than uranium-235. These numbers don't reflect biological damage. Plutonium is harmful only if it is ingested or inhaled. But horses that graze on a plutonium field are more likely to ingest or inhale the radionuclides than those that don't.

It was sadly ironic that the stallion found safety from humans and their noisy vehicles in the most radioactive outdoor environment on the planet. Since Przewalski's horses have survived only in captivity, it seemed to me that their human stewards bore them greater responsibility than, say, wild animals and I wondered whether it was right to deliberately release the horses into a radioactive zone when they have no way of knowing where it's safe and where it isn't.

Nevertheless, by releasing the horses into the zone, and breeding more horses to replace them at Askania Nova, there were 62 more

Przewalski's horses on the planet than there would have been otherwise. In short, I was ambivalent about whether I thought it was a good idea.

Vypad, in any case, was more fortunate. No square kilometer in his range exceeded 4,000 becquerels of plutonium. In general, it is one of the zone's least contaminated areas. Maybe that's why the mare that died was within radiation norms. A boar shot in Vypad's range registered less than 100 becquerels per kilogram of muscle—well below permissible limits—although animals in other parts of the zone can register thousands of becquerels.

Overlooking a field of experimental rye, containing several bounding roe deer and no horses, Sasha lamented: "Look at all this great grass. Why aren't they here?"

Visibility was so poor because of the falling snow that binoculars were of little help. In the ghost village of Korohod, where our jeep got stuck in Chapter 4, Sasha climbed to the top of a fire tower for a better view while the rest of us got out to stretch our legs and breath some fresh air. It was windy outside and the snowflakes blew into my face like tiny pins.

Windy days after a stretch of dry weather are the worst times to be in the zone because the wind shears dust—and radionuclides—resuspending them into the air where they can be inhaled. It seemed counterintuitive to me, but large particles are more easily resuspended than small ones because they have a larger surface area for the wind shear to act on. Gravity quickly pulls back on particles that are larger than a few dozen micrometers, so even very high winds usually won't raise them high enough to be inhaled. Instead, while wind carries smaller particles far afield, the larger ones just creep incrementally along the surface.

A tornado could carry more than a curie of radionuclides outside of the zone, but even less dramatic weather spreads contamination, just by blowing about fallen leaves, which contain a good part of the radionuclides taken up by trees. That's why differences in radioactivity levels between the "dirtiest" and "cleanest" parts of the zone have been gradually growing smaller in the years since the disaster. Clean leaves blow into dirty sections, diluting the contamination, while radioactive leaves skip into clean areas, increasing it. The contamination is still quite patchy, however. Radionuclides in grass samples taken from the same field can differ by a factor of two. The same is not true of trees, however, because their extensive root systems negate the patchiness.

Rain can also spread surface radioactivity, especially the first rain-drops falling on very dry ground and raising tiny puffs of dust. But a lot of rain tends to keep radionuclides that are in the ground from wandering very far, even on very windy days. Wet soil particles are heavy and sticky, so the wind's transport of radionuclides on rainy days is nearly zero. Snow cover also keeps them in their place, though the light snow falling that day had barely started to actually cover anything so it probably offered little protection to us.

While windy days usually raise radiation levels in the air, not all meteorological peaks come from stirring up radionuclides that are in the ground. Scientists were baffled when radiation levels in Europe surged at the end of May 1998. There is some evidence that the source was a Spanish steel mill smelting radioactive scrap metal. But the radionuclides may, in fact, have rained down from the stratosphere, where some radionuclides from atmospheric nuclear testing and Chernobyl ended up. Although it is generally aloof from the tropo-sphere—where we live and where our weather happens—sudden changes in wind direction can cause the stratosphere to leak some of its contents towards Earth. That may be what happened in the anoma-lous May of 1998.

There have been other strange peaks as well. One that occurred in April 2002 may have been due to a radioactive release from the Sar-cophagus or one of the facilities for processing nuclear waste.

After Sasha climbed down the fire tower bearing no good news, we climbed back into the jeep and he directed the driver over the zone's labyrinthine firebreaks, trails, and fields with the confidence (astonish-ing to me) of knowing exactly where he was going in the wilderness. The going was rough at times, like violent turbulence with no seat belts. On several occasions my head nearly banged against the roof.

I wondered how much radioactivity we were kicking up in our wake. Any kind of mechanical work on contaminated territory can ex-pose the more radionuclide-rich layers of soil beneath the surface and locally pick up dust that could get carried off a distance by the wind. That mechanical work could be accomplished by the jeep's wheels, our shoes, or the hoofs of some galloping Przewalski's horses.

Because Przewalski's herds need such big spaces, the number of herds will be limited naturally by the zone's size. Tatiana calculated that its 200,000-plus hectares can eventually accommodate from 1,000 to 2,000 horses. But like most zone scientists, her imagination often seemed to end at the international border. Belarus had an equivalent, if slightly smaller, exclusion zone and the border meant nothing to animals—like the bear that evidently wandered between the two countries in 2003—looking for new home ranges.

"The actual number may be smaller, because some young stallions may leave the zone in search of mares and may mate with domestic horses," she explained.

A woman I know once took a morning stroll through the town of Chornobyl only to come across a lone Przewalski's stallion grazing by the town's vintage Lenin monument.

"But that's a problem in all Przewalski's introduction programs. In Mongolia, they capture the stallions and bring them back to the reserve. In Hungary, they've decided to limit population growth with contraceptives. As for what to do here . . . ," Tatiana sighed. "We don't know."

In a young forest of birches and pines inside the 10-kilometer zone, we spotted a pair of moose just a dozen yards or so away. Since it was still early in the season for the males to drop their antlers, the absence of antlers meant that they were female. One cow was resting on the ground and barely moved as we drove by, but her companion loped a few strides deeper into the woods before stopping and turning her head to watch us. I had never seen a moose so close—the bull I saw in Belarus was some distance away—and was stunned by their huge size.

Although Ukraine tried to maintain the barbed wire fence surrounding its portion of the 30-kilometer zone, it was more than 100 miles long and was down in many places. Belarus, meanwhile, didn't even have a fence around its radioactive nature reserve. So there is little to prevent wild animals from passing freely in and out.

As with all zone animals that wander outside its borders, the Przewalski's bachelors were at risk from poachers. "But these are very, very intelligent animals," Tatiana said. "And when they perceive the difference between their quiet life here and the stresses outside the zone, they'll quickly return."

It was strange to think of a radioactive zone as a quiet sanctuary for any animals, much less endangered and exotic species like Przewalski's horses. Actually, however, the Przewalski's horses' life in the zone may not remain quiet if wolves figure out how to hunt them. "Some specialists told us that it can take wolves three years to learn to hunt new prey," said Tatiana. "But five years have passed and the wolves evidently still haven't learned to hunt the horses." One reason may be that they have no need to. The zone is teeming with other meals for wolves.

"This is where we saw a pack of 12 wolves trying to chase down a lone stallion," said Tatiana, as we drove through a fallow field succumbing to a future forest of pine saplings. "But he kicked one of the wolves in the teeth and got away easily."

While watching a small group of roe deer canter along the forest in the distance, I thought I saw a large black object moving near some bushes about a hundred feet away.

"Isn't that a moose?" I asked, pointing out my window. Everyone turned to look but didn't see anything. "Just bushes," Sasha started to say and I felt slightly foolish for imagining things until he exclaimed: "Boars!"

Woolly pigs started emerging from the brush and galloping parallel to our path, their dark, chunky bodies clearly visible against the snow, looking like miniature bison.

"I've never seen them so plainly before," said Sasha, counting them quickly. "There must be 25 of them!" Feeling vindicated, I did my own rough estimate while realizing yet again why wild animal counts can only be approximate. They were moving too fast to count individually, so I estimated the space occupied by five boars and then tried to count how many of those spaces the herd occupied. "Thirty," I said.

"Forty," said Natalia as more wild pigs emerged. Chuckling over our escalating estimates, Tatiana said: "It'll be 100 by the time we get back to Chornobyl." But after the boars ran off, everyone agreed that it was an exceptionally large herd. "There were a lot of young ones, too," Tatiana observed.

"Not a lot if you consider that each sow gives birth to six or so piglets," said Sasha. "A lot of them die."

"If all those piglets survived, we'd be drowning in boars," said Natalia, adding: "So far, we've seen all of the zone's ungulate species—except the horses."

"Wolves like to hunt boars because each boar only protects itself," Tatiana explained. "Horses protect themselves as well as the entire herd. Herd stallions are extraordinarily protective and strong animals."

Once, when the scientists rode out to Vypad's herd in a horse-drawn wagon, the stallion tried to attack the cart horse but didn't actually go through with it only because there were people in the cart. "We try not to tame them, so that they don't consider people too familiar." said Tatiana. "Our horses almost never act aggressively towards people. But if we hadn't been in the wagon, Vypad would have probably destroyed it and the horse."

Once, a new stud stallion that Askania Nova brought from Slovakia attacked a horse carrying a rider and had to be beaten off with sticks and clubs. He was from a zoo and didn't fear people at all.

A different complaint of some zoo breeders is a tendency to exclude aggressive Przewalski's horses from breeding. "They think this reduces the horses' chances for survival in the wild," said Tatiana. "But at Askania Nova, we think that a very aggressive animal *won't* survive in the wild. If a stallion keeps chasing and attacking wolves, at some point they'll get him. The stallion has to be smart, and control the herd, getting it to a safe place instead of attacking predators—whether they are wolves or humans. He should attack the predator only if the predator actually attacks the herd."

Soon we came to Cherevach, a former village on a deep-orange patch it shared with Chornobyl on the radioactive cesium maps. Sasha spotted fresh horse tracks and steaming manure and told Mykola to turn right and drive across a sandy field of rolling moraines formed by glaciers when Przewalski's horses first roamed the region.

Five hours had passed. With no sign of the Przewalski's, it was turning into a frustrating wild horse chase for the scientists—though I was having a fine time watching the other animals that crossed our path with exciting regularity. By lunchtime we had counted 4 red deer, 5 moose, 15 roe deer, and about 40 boars in a relatively small patch of what most of the world considered a dead zone.

I had planned to leave the zone by 2 p.m. to get back to Kiev before it was dark. Outside of cities and towns, Ukraine's roads are unlit, and driving on them in a remote no-man's-land was a little too adventurous for me, especially when it was freezing and the roads hadn't been cleared of the day's snow. But when we still hadn't found the horses by my planned departure time, I decided to stay another night. I hadn't jolted around in a lurching jeep for five hours just to give up so easily.

"If it was a hot summer day, the horses would just stand around in the same place," said Natalia, scanning the horizon as we drove. "But in bad weather, they run around more."

Our failure to see the horses reminded me of a conviction I entertained in third grade after a class trip to the New York Hall of Science. To this day, I remember the terrarium with a small sign reading: "The Impact of Radiation on Rats." There was nothing in it except some plants, which convinced me that radiation had made the rats invisible! I'm not sure why I kept that stunning observation to myself. Maybe because I concluded that it was something everyone knew and I would look foolish commenting on the obvious. Certainly, none of my classmates seemed surprised by the invisible radioactive rats.

When we came to a wall of forest standing between us and what Sasha considered the right way to go, Mykola just gunned the motor and crashed right in, crushing saplings, small trees, bushes, and fallen logs and scaring off two roe deer that scampered into the forest. Then he descended onto a relatively smooth, sandy path that he drove down with the utter confidence of someone who knows that he won't be running into any oncoming traffic, anywhere, anytime soon.

He stopped when Sasha spotted a fresh animal skeleton in a deep pile of snow-dusted grass on the side of the trail, but it was only a spine and ribcage. The head was missing, as were the legs, so there were no hoofs that would have helped identify it. From its size, the scientists guessed that it was a roe deer or a boar.

BORN FREE

After driving for nearly eight hours over 60 miles of labyrinthine trails in Chernobyl back country, Sasha finally spotted the horses grazing on the edge of a forest in Korohod—the same village where he climbed the fire tower. But getting to them took half an hour, 20 minutes of which we spent crossing about 100 yards that had been so deeply gutted by boars that the jeep could move only a foot at a time, plunging into each crater and then emerging. Plunging and emerging. What I had thought were rough roads seemed like interstates in comparison. It was like riding a horse that was bucking and rearing in slow motion.

"Now, imagine if all the boars' piglets actually survived," Tatiana commented dryly after a Richter-scale plunge. "The entire zone would be like this."

In the early postdisaster years, when radioactivity was on the surface, boars' plowing was one of nature's ways of fixing the stuff deeper into the soil. But now that 95 percent of the radionuclides are in the top two inches of the ground, the boars digging can expose them to dispersal by wind.

It was just five days short of being the shortest day of the year and the light was fading fast when we finally reached the horses. The snow had stopped, but the skies were gloomy.

"Won't the sound of the jeep scare them away?" I asked when we stopped in view of them, but not too close, and piled out. It was windy and freezing outside, and Mykola stayed inside the jeep and kept the engine running.

"Nah," said Sasha, opening the back of the jeep to take out a large burlap bag of oats. "For them, the sound of the engine is the sound of food."

Indeed, the horses started approaching as soon as we stopped. Though they had spread out to graze, the horses herded together along the way and stopped about 15 yards from us. Sasha approached them slowly with the bag and dumped the grain into five small piles separated by a short distance.

"A quick way of determining the herd's hierarchy is with supplemental feeding," Tatiana explained. "You pour about two kilograms into each pile and separate the piles by a meter or two. They shouldn't be so close together that one horse can take it all. But they should be close enough for the horses to have to fight for the food. As a rule, a horse that chases everyone away is higher in the hierarchy."

I asked how the Chernobyl horse's condition compares to the Przewalski's horses in Askania Nova and was surprised to learn that they were actually better fed.

"Askania Nova is in the steppe. Except for springtime, the main fodder is dried grasses," Natalia explained. "But green pastures are more nutritious, and in Polissia, which has a much wetter climate, there's green fodder from spring to fall."

Indeed, the horses had diversified their diet in the new surroundings. Aside from grass, they also browsed on birch and dog-rose twigs, even pine needles.

"We were shocked," said Natalia. "Pine needles are considered poisonous for domestic horses."

But the Przewalski's horses were evidently doing fine eating pine, and radioactive pine at that.

The horses divided into groups of uneven and shifting size, giving the scientists an opportunity—fading quickly with the daylight—to count and identify the horses, figure out who had foaled, and record each youngster's sex, color, and markings.

About as tall as large ponies but more robust, most of the Przewalski's horses ranged in color from hay to chestnut, though they looked very much alike to me on such short acquaintance and lacked the range of colors and markings you see on domestic horses. All of them had black stockings that faded to stripes on the backs of their legs, creamy white muzzles with black nostrils, and stand-up manes like zebras, with no forelock.

They also displayed the distinctive Przewalski's tail. On domestic horses, all of the tail hairs grow long and are the same color. But the Przewalski's tail is two-toned and two-layered. The top, or dock, layer is short and the same color as the horse's body, while the bottom layer is long and black. Some horses looked as though someone had teased their dock hairs into a beehive hairstyle.

The most distinctive of all the horses was Vypad, with his dark chocolate pelage and powerful build. Przewalski's stallions' necks grow in girth as they age, and 15-year-old Vypad's was like a Clydesdale's.

"In Askania, Vypad was in the bachelor group. Though he's a fine specimen, we have to strictly limit breeding because we have no room for more horses," Tatiana explained. "But when he turned 10, he started escaping from the bachelors' enclosure, jumping over six-foot metal fences, to run into the steppe where the mares were. He had to be captured and brought back, but kept escaping nearly every day. So, when the time came to choose horses to send here to Chernobyl, we decided that a stallion with that kind of character was an excellent choice to lead a herd. And we were right."

Vypad was munching from a grain pile on the edge of the herd, together with a pretty hay-colored mare with a creamy belly, blond highlights in her mane, and eyes ringed with gold. She was the lightest horse in the herd and almost as distinctive as Vypad. The scientists identified her officially as No. 2366—her number in the international Przewalski's studbook—and nicknamed her Vyzitka, or "visiting card." I thought of her as Blondie (Plate 5).

The studbook was created in 1959 by the Prague Zoo, when the number of Przewalski's horses in captivity was low, the numbers in the wild were still lower, and the dangers of inbreeding were high. Today, it is a unique record of one of the most intensely managed exotic species in captivity.

"That's the lowest-ranking mare," said Tatiana, pointing to Blondie. "She's staying close to Vypad because none of the other mares dare to pick on her while he's nearby. If one of them tried, he'd wallop her. He doesn't tolerate aggression in his presence."

But there was aggression going on in places where he wasn't looking. While a few foals nursed and other horses engaged in placid mutual grooming, the rest milled around, bumping about their hierarchy in sudden eruptions of spins and kicks, flattened ears, and snapping mouths.

Przewalski's herds behave in much the same way as domestic horse herds, such as American mustangs, when there is little human interference. Yet because the existing Przewalski's population has been kept in captivity for up to 12 generations, it is difficult to know if the similarities exist because Przewalski's and domestic horses have a close evolutionary relationship or because both have lived in captivity.

Tatiana explained that there is no way to diagram the herd's social structure in such a short time. It required many hours of observation, especially since the hierarchy wasn't linear but resembled a network.

"If one horse submits to another's warning once, it doesn't mean much. But if it submits five times, you can make conclusions about ranking," she said. "Sometimes submitting is as subtle as just stepping aside when a higher-ranking horse passes."

With the birth of young colts and fillies, Blondie finally had even lower-ranking individuals to pick on and did so with great evident enjoyment, although her low ranking with the adult mares was unchanged.

Like all of Askania Nova's Przewalski's horses, the Chernobyl herds are descended from Orlitza III, the last of the species caught in the wild. Brought to Askania Nova in 1957, Orlitza III became enormously important in the Przewalski's captive breeding programs. After World War II, when many of the horses died, the 31 horses that survived were all descendants of 12 founders, one of whom was the domestic mare foster mother. Rates of inbreeding were so high as to lower fitness, fertility, and life span. By injecting desperately needed new blood into the Przewalski's pedigree, Orlitza III did much to improve the captive population, including its fertility. Orlitza's descendants produced more foals than other Przewalski's lines. Perhaps Vypad inherited his strong need to breed from his great-great grand dam.

Next to Vypad and Blondie, a lone mare was trying to keep a pile all to herself. If any other horse tried to approach it, she'd flatten her ears and snap, or whirl and threaten to kick. I didn't see her actually connect with any of her would-be dining partners, but they got the message and generally wandered over to Vypad's end.

"She must be pretty high in the hierarchy," I observed.

"Actually, lower-ranked horses can also be aggressive," said Tatiana. "You don't just look at whether the horse displays aggression, but at the result. If one horse threatens to bite another to move it away from the grain pile, but the other horse doesn't move, the first one is probably lower ranked."

Tatiana returned her attention to identifying the horses in the remaining minutes before darkness fell. "If only we had found them half an hour earlier," she lamented.

Once it was too dark to identify the horses, I followed Tatiana and Natalia to collect manure samples so that they could check the horse's internal parasites.

"Stay close to us so that the horses don't think we're circling them," said Natalia. "They'll consider that threatening."

As soon as we started approaching, the horses herded together and moved away, but they were clearly reluctant to leave the remaining oats and didn't go very far.

Blanketed with dried grasses and light snow, the ground was rutted by boars, and the light was fading very fast, making it hard for me to distinguish dung from dirt. Not being of much use, I trudged back to the jeep to sit in a relatively warm place for a few minutes. It was bitterly cold. But the hardy Przewalski's horses seemed untroubled by the temperature. Horses handle cold weather well when it's dry, but they don't much like rain or wind.

When I could feel my fingertips again, the horses had returned to the grain piles and the scientists were returning to the car, their feet loudly crunching the dried grass. Sasha was trying to shush them so that he could better hear a distant barking.

"Lynx," said Sasha. "Maybe red deer."

"It's too loud to be a deer," said Mykola.

When the barking stopped, Tatiana chided me: "Mary, why did you leave? Didn't you see what happened? They came to us and stood just a few feet away."

From her excitement it was clear I had missed something extraordinary. It was also clear that these two women cared for the horses too much to intentionally do anything to harm them. It was hard not to share their optimism about the Przewalski's horses' future in the zone.

"It's when we crouched to collect the samples," Natalia explained. "We seemed smaller to them—less threatening—and they also probably thought we were bringing more oats."

It was the first time that the scientists had been so close to the horses.

"They're very curious," said Tatiana after we had piled back into the jeep and started our return journey to Chornobyl. "If you want to photograph them, walk towards them swiftly and then suddenly crouch or sit down. When you seem that small, they'll come over to see what's going on. But that's only with Vypad's herd. Since we work with them so closely, they aren't that scared of us."

"If you tried that with Volny's herd, they'd hightail it out of there," Natalia added. Volny was much shyer than Vypad.

The scientists counted 25 horses in Vypad's family: 8 adult mares that had come from Askania Nova together with 7 foals and 9 juveniles born in the wild. The stallion was a fertile fellow and had had a prolific year. But they failed to identify all of the foals because larger horses often blocked their view. Definitely identifying the juveniles was also tricky since some looked very much alike and had to be observed at different angles.

"It would be easy if they just stood still," said Natalia. "But they don't."

"For example, we think that the very aggressive mare was the first Chernobyl foal born in 1999," Tatiana added. "But we're not completely sure."

That would make the young mare, nicknamed Pripyat, more than four years old and well past the age when stallions chase their daughters out of the herd. To prevent inbreeding, juveniles are expelled when they are between two and three years old. That year, Vypad had already expelled some of his older offspring, and the scientists were eagerly

PLATE 1 The ruined fourth reactor.

PLATE 2 The 10-kilometer zone in February 1987.

PLATE 3 A Red Forest pine tree.

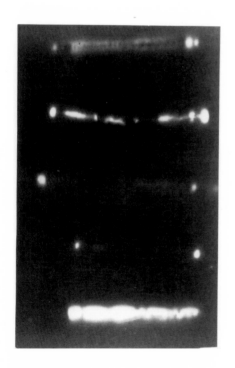

PLATE 4 Hot particles on a
pine needle.

PLATE 5   Blondie and Vypad.

PLATE 6 Maria Shevchenko and Oleksandr Tkachenko.

PLATE 7 The Shelter Object.

PLATE 8 The elephant's foot.

hoping that some bachelor stallions would pick up the young mares to form new harem groups and increase the Chernobyl Przewalski's horses' genetic diversity.

If the mare was Pripyat, perhaps Vypad was letting her stay because he wasn't her father. Her mother got pregnant in Askania Nova, where Vypad wasn't allowed to breed. Actually, some Przewalski's stallions have been known to kill foals when they take over a harem. But for all his fearsome appearance and reputation, Vypad was a gentle daddy. He was rarely aggressive towards the foals, and in the winter, when the horses dig pits in the snow to reach the grass beneath, he shared his pit with the foals. No mare in the herd has been seen sharing her snow pit with a foal that isn't her own. Of course, each mare has but one foal at a time. Vypad had to take care of all his offspring.

IN THE SHADOW OF CHERNOBYL-2

It was still dark outside when the jeep picked me up at the Chornobyl hotel at seven the next morning. Eager to continue their study of Vypad's herd, the scientists wanted to take full advantage of the short winter day and, hopefully, catch the horses at the same place we had seen them the previous evening, before they galloped off to a different part of their enormous range. In fact they were so eager, they kept dozing off in the back seat because, Tatiana later confessed, they had been so worried their alarm wouldn't go off in the morning, they couldn't sleep half the night.

The jeep's headlights illuminated a path on the snow-packed roads and trails as we drove 40 minutes through the darkness to the field in Korohod, only to find that the horses had already left. The jeep stank of exhaust even worse than I remembered, so it was a relief to get out while Sasha and the scientists examined the hoof prints in the snow to see which way the herd had gone. The rising sun was merely a pink smear above the tree line when we set off in the direction of the 10-kilometer zone, back on one of the paths we had taken the previous day. I recognized it only because of an empty sardine can someone had tossed aside in the sand.

After another 40 minutes of bumping along rough trails and then crossing another cratered field dug by boars, we found the horses near the border of the 10-kilometer zone and in view of the abandoned Soviet radar station nicknamed Chernobyl-2. It would be redundant to say that the station had been secret in Soviet times. Nearly anything that wasn't in plain sight was a Soviet secret, and to make sure that too many people didn't see what was in plain sight, permits were required to enter entire cities and regions. Foreigners were banned altogether from many places.

The sun had risen over overcast skies when we stopped, but the freezing, windy weather persisted. Once again, Sasha poured grain into separate piles and the horses approached, standing less than 30 feet away from us as they divided into groups around the grain piles and repeated their hierarchy rituals. The aggressive mare of yesterday was evidently on better behavior, though, making her harder to identify.

The scientists set to work immediately, examining the horses, making notes, and holding esoteric exchanges along the lines of:

"Foal nursing number 161. Male. Dark spot on nose."
"That's not 161. She has a star. It's 159."
"Damn! 163 just blocked my view."

Though all the horses that came from the nature reserve have international studbook numbers, the scientists were using Askania Nova's internal nomenclature that they use to count only their own animals. They were recording the foals' markings for future identification and trying to figure out which foals belonged to which dams. One way to be reasonably certain was to see which mare was nursing them. Even that wasn't a sure thing, though. In zoos, Przewalski's dams have been known to nurse their daughters' foals, and daughters have been observed nursing their mothers' foals. But since sexually mature offspring are expelled under natural conditions, that shouldn't happen in the wild.

While the horses engaged in their hierarchical rituals of snapping, spinning, and kicking over access to the tasty treats, one yearling colt evidently decided that fighting over swiftly dwindling piles wasn't worth the trouble when he caught a whiff of the large sack of remaining oats that Sasha had placed behind the jeep.

"He's the tamest," said Natalia as the colt faced the jeep from a distance of about 20 feet and stretched his neck to smell the oats. "He's the least afraid of people."

A few other horses gathered near him as he approached the sack incrementally. Step, stop. Step, stop. If another horse got too close, he whirled and kicked, but since he was clearly the only one among them who was brave enough to lead the way to the sack, the rest followed meekly enough and kept a respectful distance. By the time he was a nose away from the jeep, about a quarter of the herd was right behind him, including Vypad who didn't seem to be pulling rank in any way. But when the colt finally touched his nose to the sack, his own audacity—or our proximity just a few yards away—proved to be too much and he spun around, prompting all the horses to gallop to a safer distance.

The horses had finished eating the oats by then and Sasha wanted to pour them more, but Tatiana stopped him.

"They're packed too close together by the piles and we can't get a good look at them," she explained. "If there's no more food here, they'll start grazing and spread out."

Instead, the horses just stood and watched us—and the sack—expectantly. Mykola, the driver, sat on it to keep the horses from approaching it again, but this didn't keep them from hoping, even after Mykola put the sack back in the jeep and turned on the engine and clouds of exhaust fumes floated over the herd.

"So long as they see the jeep, they don't lose hope," said Natalia.

Blondie stood by herself, a short distance from the rest of the herd, perhaps because higher-ranking mares had surrounded Vypad. He gently nuzzled a golden mare's withers and rubbed his face against hers. It seemed like a tender display except it was actually a sign of dominance. The dominant animal rubs but does not allow itself to be rubbed.

But the display angered the alpha mare, who charged up to the pair, snapped to chase her rival away and then opened her mouth to stroke the stallion all over his head. Thus assured of his attention, she ambled away.

Though Vypad was the highest-ranking horse in his herd, the stallion is not always the alpha animal in harem groups, especially in captivity. If a young stallion is joined to an established group of mares as a stud, high-ranking (and generally older) mares simply won't breed with him. A stallion must outrank a mare before he can mate with her.

In the wild, though, new herds are formed when bachelor stallions fight for access to young mares that have been expelled from their family harems. The winning stallions are almost always older and more experienced than their rivals and more dominant than their first mares, so that the stallion is top ranking. But when a strong and experienced bachelor eventually wins Vypad's herd—and that day will inevitably come—it is unlikely that he'll be able to dominate, or breed with, the older mares.

After about an hour, the horses decided that no more food was forthcoming and they started slowly moving a short distance away from us to graze. Vypad followed in the rear, stopping occasionally to turn around and look at the jeep.

When the horses spread out, Sasha and the scientists approached them hoping to finally get a good look at the foals, while I climbed into the jeep to warm up and make some notes. The smell of exhaust inside was suffocating, though. After a few moments, I climbed back out into the cold, upwind from the exhaust belching from the jeep, and walked slowly towards the horses, trying to imagine myself an Ice Age huntress scouting a horse herd on the periglacial steppe that covered Polissia when modern humans arrived in the region around 25,000 to 30,000 years ago.

After a while, the horses had grazed their way some distance from the jeep, so Mykola started the engine to drive closer. But for the horses, the engine meant "grain!" and they quickly herded together, prompting annoyed grumbling from the scientists when they returned to the car, with the herd following close behind.

For a few moments, horses and humans stood quietly and watched each other, filled with hopes of getting something. The horses' hopes were simple: food. But what were we humans hoping for? I couldn't speak for anyone, but I was hoping for a few more moments of communion with animals that seemed to symbolize our enormously complex and complicated relationship with our planet. Humans brought Przewalski's horses to the brink of extinction and then gave the species a chance for a new future in the wild. How ironic it was that one of the few places on Earth where they can live in the wild, free from human threats, is a radioactive land that threatens humans more than it seems to threaten them.

6

Wormwood Waters

. . . and the third part of the waters became wormwood; and
many men died of the waters because they were made bitter.

Revelation 8:11

I n the first months after the Chernobyl disaster, all of the zone's
waterways were metaphorically bitter with radiation. It flowed
through the entire aquatic cycle: falling with the rain; filtering
through the radioactive cloud that meandered about the western Soviet Union for months; trickling across contaminated surfaces coated
with fallout; washing out in the tons and tons of water and chemicals
used to decontaminate towns, villages, and the nuclear plant's grounds.

Winds, water currents, and waves spread the radioactive surface
film on rivers and lakes to the banks, creating so-called riverbank
anomalies—highly contaminated strips of soil that were at the level of
the radionuclide-coated water at the time of the disaster. They were
detected in all zone waterways and along the Dnieper basin, from
Belarus to the Black Sea.

In the immediate aftermath, one liter of water in the nuclear plant's
cooling pond measured tens of thousands of becquerels of radioactivity, much of it from short-lived isotopes such as iodine-131, barium-
140, and tellurium-132 that covered the surface and slowly sank into
its depths. But the levels dropped quickly as these short-lived radionuclides decayed and longer-lived isotopes and fuel particles sank into
the bottom sediments, although they spiked occasionally when radionuclides were washed out by cleanup works.

As time passed, 95 percent of the radioactivity ended up on the bottom. In closed aquatic systems such as the cooling pond, nuclides and fuel particles penetrated up to 10 centimeters (four inches) into the bottom sediments. Although fuel particles disintegrate more slowly in bottom sediments than in soil, they do cause continuous secondary contamination of water.

In the Pripyat River, bottom sediments nearest the reactor were so contaminated in 1988 that they were classified as solid radioactive waste. But flowing water systems such as rivers gradually dilute contaminants by transporting them downstream. In general, the higher the water flow, the greater the dilution and the lower the contamination. As one result, radioactivity levels in the zone's rivers fell significantly. A decade after the disaster, most river bottoms in the 30-kilometer zone were virtually clean, measuring tens to hundreds of becquerels per kilogram and—in exceptional cases—in the thousands.

Yet the cleansing of some zone waterways by flowing water has a downside: nearly all of the radionuclides that get out of the zone, contaminating larger and larger regions, travel with water.

THE POLDER

The deep crow's feet around Marian Sikora's pale blue eyes made him look about 50, though only a little gray sprinkled his brown hair and mustache. A hydrology technician by training, he was deputy director at the tongue-twisting *Chernobylvodexpluatatsia* agency. That translates roughly as Chernobyl Water Exploitation and reflects a still-Soviet habit of naming things by running the words and syllables together into single unwieldy neologisms. For the uninitiated, the results were slightly more informative than acronyms, but the ridiculously long words often didn't even save much space on the signage. *Chernobylvodexpluatatsia* was too cumbersome for anyone to use regularly. The people who worked there referred to it as CheVE, as did I.

I was riding with Sikora in an old CheVE van on the left bank of the Pripyat River. Although it is a major river for Belarus, the 440-mile (710-kilometer) Pripyat is not very significant in Ukraine, which has much larger and economically significant rivers such as the Dnieper and Dniester. During more than a dozen trips to the Ukrainian zone, I had always stayed on the right bank of the river—except when I crossed it to go to Belarus, but that was very brief. The town of Chornobyl was on the right bank as were the power plant, the town of Pripyat, the equipment graveyards, and just about anything of interest to most visitors. But on this trip I wanted to see something different and the only place to see it—indeed the only place to see anything remotely like it in the world—was across the river from the nuclear power plant.

The darkest colors on the radiation maps marred the spot, including a brown smudge near the village of Krasne that is one of the horns of dense debris from the April 26 explosion. But unlike the other horn, which killed the Red Forest but at least stays put, the right bank horn belonged to a mobile devil that is and will be a persistent threat to the Dnieper River. The reason is that the concentrated contamination lies directly on the Pripyat River's floodplain. The entire eight-square-mile area contains an estimated total of 10,000 curies of cesium-137 and 6,500 curies of strontium-90. When it's flooded, the radionuclides leach into the river, which spills them into the Kiev Sea, whence they gradually migrate down the cascade of Dnieper reservoirs that supply drinking water for 20 million people and irrigate nearly 2 million hectares of land in the arid southern steppe.

In 1991, when ice jams near the power plant caused intense flooding of the riverbanks in the 10-kilometer zone, strontium levels in the water reached as high as 12,000 becquerels in a cubic meter of water! Maximum allowable levels in Ukraine are 2,000 becquerels, though actual concentrations in the Kiev water supply are usually in double digits.

It took about 40 days for the contaminated water to travel from the northern part of the Kiev Sea closest to Chernobyl to the southern portion nearest the Ukrainian capital. By that time, the radioactivity had diluted to officially safe levels. But they were still four times higher than normal and caused a lot of anxiety—for me, too, because I had moved to Kiev a month or so earlier.

The ice jams lasted for more than a month and eventually had to be broken up by bombing. When the flooding ended, the river had carried more than 100 curies of strontium into the Dnieper system and the vast majority of it came from the Krasne floodplain.

Something clearly had to be done to keep the floodplain dry. So, after the floods ended, the Ukrainian government transformed the area into a polder. "Polder" is a Dutch word for a parcel of low-lying land reclaimed from the sea, a river, or another body of water and surrounded by dikes and canals that regulate water levels. More than a third of Holland is below sea level, so polders make up a good portion of the country—including the biggest cities and best farmland.

The Chernobyl polder was, territorially, more modest, but its creation was the most important measure taken to keep radionuclides out of a major Ukrainian water supply. This, at least, was what I had read in numerous articles in Ukrainian scientific journals. The trouble was that none of them described the thing or explained how it worked. An Internet search for "Chernobyl polder" in English produced 129 results. The same search in Ukrainian produced only one result, and in Russian it produced zero. None of the results were useful. So, Sikora was taking me to see it firsthand.

Cleanup barges and ferries rusted in a bay of the Pripyat River under the bridge from Chornobyl. The river was gray under a steely and cool October sky. Some trees in Kiev were still green, but 50 miles to the north the foliage had already turned, though it lacked vibrancy under the dull clouds.

In a few minutes we were at the Paryshiv checkpoint. Actually, its official name was Paryshiv-2, and it was different from the Paryshiv-1 checkpoint I crossed on the border of the Ukrainian zone when traveling to Belarus. Paryshiv-2 was located less than two miles from Paryshiv-1 and was actually a very strange checkpoint because it wasn't on any zone border and it was nowhere near the nuclear plant. In fact, it was impossible to tell what it controlled. But Paryshiv-2 was a multipurpose checkpoint that secured an intersection. The road we were on led to and from Chornobyl. A second road led southeast, towards the Kiev Sea. Another led northeast to Paryshiv-1 and Belarus.

We wanted to turn onto the fourth road that led northwest to the 10-kilometer zone. Because Paryshiv-2 controlled a four-way intersection, each arm of the intersection was blocked by a candy-striped gate. But the guardhouse stood quite a distance away from the actual crossroads, and the guard had to walk about five minutes to open any of the gates. Compared to the pretty flower-embroidered cottages at the checkpoints in the Belarusian radiological reserve, the Ukrainian post was an eyesore that looked as though it came from a prison camp.

It took more than an hour to drive 12 miles over the paved but potholed road leading past disintegrating villages that once tilled the Pripyat floodplains. Disrobing autumn trees revealed their ruins, dull and dark but for the occasional cottage sporting bright red creepers that flashed with brilliant color even under the gray skies. Flocks of tree sparrows spilled out of abandoned orchards like brown autumn leaves.

The bright red apples ripening in the trees evidently reminded Sikora of a joke, which he proceeded to tell us:

A babushka at the market shouts: "Chernobyl apples! Get your Chernobyl apples!"

"Don't say that they're Chernobyl apples," says a passerby. "No one will buy them."

"Sure, people buy them," says the babushka, "for their husband, their wife, their mother-in-law."

There was no checkpoint at the border of the 10-kilometer zone itself and not even a sign of where it actually began. My homemade carto-graphical creation comprising three cellophane-taped pieces of Belarusian and Ukrainian topographical maps had a 10-kilometer circle that I drew with a compass. On the ground, though, the zones had very irregular borders. A length of the "ten's" barbed wire fencing ran alongside the village of Koshivka, although that was outside the 10-kilometer zone's borders on my map. Even my guides weren't sure exactly where the inner zone began.

After passing the only other vehicle we would see on the left bank that day in more than three hours of driving—a CheVE truck in the village of Zymovyshche—our driver turned left onto a dirt path run-ning alongside the railroad tracks that connect the town of Slavutich, built to replace Pripyat, with the nuclear plant. The path tilted down a bit towards the marshes that ran parallel to us, and a few times I had the feeling that we were going to tip over. But we just bumped and bounced along for half a mile and then turned onto a narrow sand dike.

"This old dike was built before the disaster to protect the riverside villages from spring floods," Sikora said, pointing it out on my map with a stubby finger. "It forms the northern border of the polder."

A canal on our right ran parallel to the dike's northern slope and collected floodwaters that splashed up and seeped through its sand barrier. Known as MK-7, the canal followed the four-mile dike from the border with Belarus and drained into the Pripyat River. Some Belarus canals drained highly contaminated water into MK-7, too, and the Ukrainian side had come up with a plan to divert those canals in a different direction so that they would spill their water much farther downstream, after the radionuclides settled out of the water and into the canal's sediments.

The dike we were on was part of a land reclamation system that crisscrossed the left bank with channels and ditches for draining the peat swamps and bogs that had once dominated the area. Put into service in 1978, the system was declared unnecessary when the zone was evacuated and abandoned.

With time, however, the channels became clogged with contaminated silt and plants, eventually backing up and flooding nearly 5,000 acres of forest—far more than had been predicted. Highly contaminated with up to 100 curies of strontium and 200 curies of cesium per square kilometer, the re-created wetlands spilled radionuclides into the Pripyat and also impeded firefighting since fire trucks couldn't get through the swamps. In 2001 the zone administration decided to renew the system, cleaning the radioactive muck out of miles of drainage ditches so that water levels on the floodplain could be controlled as needed. It was a balancing act. The forests had to be drained at the same time that the peat lands had to be kept relatively moist to reduce their risk of potentially long-lived and radioactive fires.

While my digital recorder captured Sikora's words, I gazed out the van's window and spotted a bull moose about 20 yards away, loping up the steep banks that led from the drainage channel into a copse of young birches, whose golden leaves trembled in a light breeze.

"*Los!*" I announced, using the Slavic word.

"What a beauty!" said Leonid, the driver. Like most drivers I'd met in the zone, I knew almost nothing about him and mostly saw only the back of his hair, which was blond. But I did learn that he was a hunter, skilled at spotting animals and birds.

Soon afterwards, he pointed out a black grouse that landed in a forest clearing near the road, and when a flock of partridges emerged from the brush and scurried on the path ahead of us, seemingly unworried and unhurried by our rattling van, Leonid obligingly slowed down and let them go at their own speed until they veered into a meadow.

"Hey, little fellows, take your time," he said.

"We're in no rush," added Sikora.

Whether we were or weren't in a rush was not up to me. Though I had, as always, planned what I called my "water trip" through

Chernobylinterinform, the agency had delivered me into CheVE's care, with CheVE escorts and CheVE transportation. Although they were making a special trip for me—Sikora gave no indication that he had been planning to visit the polder that day—I felt as though I was tagging along. Knowing how underfunded all zone agencies are, I later tried to at least reimburse Sikora for gasoline, but he gallantly refused to even hear of it.

He asked Leonid to stop when we came to a lake on our left. Lake Hlyboke, meaning "deep," was one of the most radioactively contaminated waterways in the zone. It was just across the river from the nuclear station, and the smokestack from the Sarcophagus rose above it on the horizon. The lake had a meandering edge bordered with reeds and rushes. Its still waters reflecting the gray skies looked like mercury.

After climbing out of the van, I turned on my dosimeter, which rapidly beeped out a reading of about 200 microroentgens an hour in the center of the dike. When I laid it down on the steep banks facing the lake, the reading shot up to 800.

"It's high because this side was facing the reactor when it exploded," Sikora explained. "Parts of the polder go even higher, into the milliroentgens."

On the dike's opposite slope, facing away from the plant, the liquid crystal display hovered around 300.

"The lake used to get flooded before the polder was built and spilled a lot of contamination," said Sikora. "Although nearly all of the radionuclides are buried deeply in the bottom sediments, flooding can churn them up."

Radioactivity in some lakes also gets stirred up by wind, circulating waters, and seasonal turnover of the warm and cool layers.

Like the Red Forest, Lake Hlyboke is not a place in which you want to spend too much time, so we piled back into the van after a few minutes and continued our drive along the dike until Leonid turned left onto what seemed a low and broad expanse of sand that hugged the Pripyat riverbank. The eponymous ghost town's deserted apartment buildings rose from the other side of the river.

"This is the new dike completed in 1992," said Sikora. "It runs along the riverbank to prevent floodwaters from entering the polder."

The new dike was seven miles long and seemed lower than the old one, but this was because the latter was narrow, with steep slopes, while the new dike was so broad that I could barely see its gradient. Both were the same height: about 23 feet above the river. But the new dike was wider to reduce seepage. A sand dike can prevent flooding, but it can't stop some of the floodwaters from filtering through the pores and spaces between sand grains. The wider the dike, the farther the seeping water must travel and much of it loses power before it can get inside the polder. A canal ran alongside the dike inside the polder to collect whatever water did seep through and also to drain rainwater from the polder, carrying it to a collecting pond, where a pumping station kicked in automatically whenever the water exceeded a certain level. Instead of spilling into the polder and rinsing out radionuclides, the excess canal waters were pumped through pipes into an old channel of the Pripyat River near the village of Krasne.

The waters in the polder canal are far from clean. Rain drains through the contaminated territory, washing out radionuclides—though not all of them equally. Although plutonium will be around for many millennia, it is not very mobile in water. Once it sinks into soil and sediments, it tends to stay there. Cesium-137, which sticks to things in the sediments, also isn't very mobile in water, although it can dissolve under acid conditions such as those in peat bogs. But in the waters of nonacidic closed lakes such as Hlyboke and the cooling pond, cesium levels have been falling over the years. The exception is the polder, where cesium levels increased in 2001 because mobile forms of the radioactive isotope were flushed into the water during earthworks on the drainage canals.

Strontium is by far the greatest concern, since it is much more mobile and its presence in water has been increasing. Indeed, strontium levels doubled between 1986 and 1994, and they were even higher during the 1991 ice jam. In 2002 the concentration of strontium in a cubic meter of the polder's waterways averaged around 15,000 becquerels and went as high as 21,000. Strontium will continue to be released as the hot fuel particles dissolve with time.

Floodwaters can also carry radionuclides, especially when bottom sediments get churned. But the nuclides sink rather quickly into the canal's bottom sediments, and only the upper layers of water, which are relatively clean, get pumped out.

The polder doesn't completely prevent radionuclides from washing out of the floodplain. It merely reduces the amounts. If the 1991 ice jams flooded out 100 curies of strontium, comparable ice jams in 1999—when Pripyat waters levels were among the highest on record and floods could have washed out as much as 400 curies—carried out only 50. Altogether, zone ecologists calculate that the polder prevented more than 540 curies of strontium from being carried out by ice jams and spring floods over the years.

Whether this is a lot or a little depends, like most aspects of Chernobyl, on your perspective. Since 1986 the river has swept more than 4,000 curies of strontium and 3,000 curies of cesium out of the zone. Most of this, however, happened in the year of the disaster, when about 1,800 curies of cesium and 750 curies of strontium were carried out. After 1988 the amount of strontium was usually double the amount of cesium because cesium started bonding to elements in the soil at the same time that disintegrating fuel particles started releasing strontium.

Moreover, although it is the largest, the Pripyat is not the zone's only river. The Uzh is an important Pripyat tributary that flows along the south of the zone while the Braginka flows through Belarus. Both empty into the mouth of the Pripyat. Since 1986 these rivers have together washed a total of more than 12,000 curies of radionuclides out of the zone, but recent years have been contributing steadily reduced shares of that total. In 2001 the three rivers carried out 96 curies. In 2002 the total was 47, and in 2003 it was slightly more than 40—the lowest in the postdisaster period.

It would seem logical that the people who get the highest radiation exposures from consuming this water are the residents of Kiev, because the Kiev reservoir is directly in the path of water flowing from Chernobyl. But actually, only a few percent of an average Kiev resident's radiation exposure comes from water. In contrast, water is responsible for 10 to 20 percent of the dose lower down the Dnieper basin because the river waters are used for irrigation and thus create a pathway for radionuclides to get into food. Nevertheless, in some country districts of Kiev, the dose from water may be much higher than average because small lakes, heavily used by the local population, were highly contaminated.

"So, this polder system has to be maintained for centuries," I commented as we trundled along the dike directly across the river from the Sarcophagus.

"If they keep cutting our budget and laying off people, it will fall apart way before that," said Sikora. "When I started at CheVE six years ago, 240 people worked there. Now there are 172."

The new dike requires constant maintenance because the river current constantly erodes the sand. Sikora said that CheVE had reinforced parts of it with gravel and rocks but didn't have enough money to complete the job. Lack of financing plagued all Chernobyl works.

As we drove off the dike, a small flock of cormorants flew out of the polder over the heads of three roe deer whose pelage was taking on its gray-brown winter cast.

Not long afterwards, Leonid spotting a wolf trotting up ahead on the road and gunned the gas with an excited whoop. It wasn't clear if he was catching up to it or chasing it, but the distinction was lost on the wolf, which scrambled ahead of us, its thick reddish tail bouncing. Just when it seemed that we were about to run right over it, Leonid screeched to a stop that jolted us all and sent my bags flying to the floor. But the wolf was unharmed, and by the time we got our bearings, it had dashed into the thicket.

"After the disaster, there were wolf-dog hybrids because the wolf population was small and there were many stray dogs left behind after the evacuation," said Leonid. "But when the wolf population stabilized, they pushed out and killed the hybrids. Wolves and dogs are competitors and don't tolerate each other."

It was the first time I had ever seen a wolf in the wild. Seeing it from inside a van suited me fine.

FISH STORY

The next day was one of those beautiful autumn days when the air is crisp, the sun bright, and the foliage brilliant. Leonid was driving Sikora and me in the 10-kilometer zone on the way to the nuclear plant. The road from Chornobyl to Chernobyl runs roughly parallel to the plant's artificial cooling pond, although it can't be seen from the road. The first evidence of its existence comes when the Sarcophagus is in view and the road curves alongside the channel that once discharged hot water from the reactor at the rate of an Olympic-sized pool every 17 seconds. That hot water cooled as it diluted, but the artificial pond's waters remained relatively warm and never froze in the winter. Many aquatic birds that normally migrated stayed there year-round. But since the last working reactor at the plant was shut down in 2000, most of the cooling pond freezes like the zone's other waterways.

We stopped near the Chernobyl nuclear power plant's parking lot, where a small bridge for its service railroad crossed the discharge channel. Since the waters had cooled, fish had started visiting the channel and the bridge was a favorite place for plant employees to toss them leftovers from lunch. Most famous were the giant Chernobyl catfish.

I carefully followed Sikora onto the bridge with the hopes of getting a glimpse of them. Between the tracks and the narrow pedestrian pavement that ran alongside, there were a lot of gaps through which I could see the murky waters below. It was not a very reassuring structure and I kept my eyes on my feet until reaching the middle of the bridge. A beam of sunlight illuminated the waters right beneath us, where dozens of normal-sized fish but no giants swam.

"There are swarms of them at lunchtime," said a disappointed Sikora as he scanned the waters.

I don't like heights or bridges—even bridges that were more solid than the one I was on—and I felt as though my glasses were in danger of slipping off when I looked straight down into the water. But when Sikora exclaimed that he saw one, I tightly clutched the handrail and looked down.

The other fish scattered when the predatory catfish entered the illuminated patch of water, its whiskers casting a shadow on the channel's concrete bottom. It had a big, blunt head and must have been at least six feet long.

Though unsuspecting visitors can occasionally be fooled into thinking that the Chernobyl catfish were gargantuan mutants, they were actually known as wels catfish (*Silurus glanis*), Europe's largest freshwater fish. Omnivorous and aggressive, wels catfish can grow to more than 18 feet in length.

As with mammals, no mutant or deformed fish have been discovered anywhere in zone waters. Scientists are stumped as to why, although it may be for the same reason that no mutant mammals have appeared either—if any mutants are born in the wild, they die. Also, the animals that received the worst genetic injuries from the disaster died before they could pass damaged genes to their offspring. Those that were born normal and survived may have been more resilient to radiation, a trait that their progeny inherited.

In fact, one study of 400 crucian carp in Chernobyl lakes found that although the fish showed many genetic changes, their appearance was completely normal. Another study of channel catfish, which are much smaller than their wels cousins, also found that they looked perfectly normal but had genetic damage such as breaks in their DNA.

The channel catfish were probably descendants of farmed fish. In the early 1980s, the cooling pond was home to one of the Soviet Union's largest industrial fish farms. Fed artificial clean feed, their radioactivity levels were actually lower than in wild fish, because wild fish eat food that still contains traces of fallout from atmospheric nuclear testing (as do we all). In fact, fish farming in the cooling pond has continued from 1987 to this day. It is experimental, and the fish can't be eaten, but their young can be grown to maturity in radioactively clean waters.

"That's a small one," said Sikora, peering down at the wels catfish under the bridge. "There are much bigger ones in there."

The catfishes' size was, in part, a result of radiation—but not because they were mutants. Since fishing is banned in the zone, the catfish can grow to their maximum length undisturbed. Although wealthy businessmen and politicians pay hefty bribes or pull strings to poach some of the zone's abundant fish or game for sport, few of them are dumb enough—or desperate enough—to catch fish in the cooling pond.

The cooling pond represents what is called a lentic aquatic system. Lentic systems, like lakes and ponds, have standing water or water with weak currents—as opposed to lotic systems, which have running water. Plants and animals in lentic systems are much more radioactive than those in lotic systems because there is no flowing water to dilute the radionuclide concentrations.

Though contamination levels in cooling pond fish have fallen well below the postdisaster high of more than 600,000 becquerels per kilogram, nearly all the fish continue to exceed radioactivity limits more than 18 years after the disaster. In contrast, less than 20 percent of Pripyat River fish are too contaminated to eat safely. But radioactivity levels in the same species of fish caught in the same place on the Pripyat River can differ from 10 to 140 times. In general, the closer fish are to the nuclear station, the higher are their radioactivity levels.

Fishes' radioactivity levels depend on what they eat and their place in the food chain since they accumulate radionuclides with their food. Little contamination gets in through their skin or gills. As a rule, cesium concentrations increase up the aquatic food chain, with predators such as catfish and pike perch registering the highest levels. But strontium levels decrease up the food chain. Herbivorous fish can contain a lot of strontium because the nuclide is bioavailable in plants, but nearly 90 percent of the strontium ends up in their bones, which don't metabolize well. So, when herbivorous fish become food for predators, their radiostrontium is not bioavailable and is excreted rather than absorbed.

The main problem in the cooling pond, however, is not its fish. Other zone waterways, such as Lake Hlyboke, are more radioactive and have more radioactive fish. The problem is that the cooling pond is artificial. Money is needed to operate the pumping stations that maintain its water levels, but since the atomic station closed down, there's no economic need to keep it going.

Back in the van, Leonid drove us around the back of the nuclear station and then turned right onto a narrow road bordered by a sandy shoulder and trees. It was the dike that surrounds the cooling pond and prevents it from spilling into the Pripyat River, since the former is about 21 feet higher than the latter. But the ever-alert Leonid had spotted what he thought was a snake on the road.

"They should already be hibernating," he said and stopped the van to hop out for a look. But he didn't actually put the hand brake on and I noticed that the landscape was moving in slow motion outside the window.

"We're moving," I said to Sikora.

"Don't worry," he assured me with an utter lack of concern. "We'll stop eventually."

We were probably moving at the pace of a slow shuffle, and the van did stop before Leonid got back in and announced that it was, indeed, a snake but a sluggish one. Still, the CheVE guys' attitude towards auto safety made me wonder about their attitudes towards radiation safety. In general, after more than a decade of living in Kiev, I was still sometimes shocked by the lackadaisical attitudes that were inherited from Soviet times and quite obviously played a role in the Chernobyl disaster.

Soon we came to a control point at the pumping station on the cooling pond's northern edge, with a view of the nuclear station's fifth and sixth reactors looming above the opposite bank. Under construction when the disaster struck in 1986, and bristling with radioactive cranes to this day, the reactors were never completed.

The small cubic guardhouse was empty. In fact, there didn't appear to be anyone around at all when we all climbed out of the van and walked past some tilted telephone poles and rusted metal structures that littered the grounds. Tall grasses grew from cracks in the concrete. Radiation levels at the center of the dike hovered around 50 microroentgens an hour.

Wind carried a light spray from the churning waters that the groaning pumps have had to pipe in from the Pripyat River since the nuclear plant was shut down and stopped discharging its hot waters into the cooling pond.

According to various estimates, the artificial pond's nine square miles contain anywhere from 450 to 7,000 curies of radioactive cesium, strontium, and transuranic elements. Most of it is in the bottom silt and concentrated in about a third of its area, near the northern portion where we were standing, although the few hundred feet closest to the pumping station are the cleanest because of the uncontaminated river water flowing in. Fish caught there display a greater variety of radioactivity levels than in other parts of the cooling pond because they include clean specimens that reside there permanently as well as visitors from more contaminated sections.

The need to maintain water levels by pumping water from the Pripyat has also introduced some river species, such as Ukrainian brook lampreys, into the pond. If the pumps stop replenishing the cooling pond with river water, much of it will eventually either evaporate or filter through the sand dikes into the Pripyat River, leaving behind a series of small radioactive ponds and exposing up to six square miles of radioactive sediments. The risks of such a development were seen in the Soviet nuclear weapons complex in 1967, when a contaminated lake in the Urals dried out during a drought and winds blew radioactive cesium, strontium, and cerium up to 75 kilometers away.

A similar situation, though on a far smaller scale, occurred at the Savannah River nuclear site in the United States, where nuclear weapons reactors spilled some radioactivity into their cooling ponds in the mid-1960s. Radioactivity levels in the ponds' flora and fauna have been declining steadily since then and were pretty much back to normal by the late 1980s. But in 1991, water levels in one of the ponds had to be lowered for repairs on a dam, temporarily exposing the contaminated sediments. As a result, some waterfowl such as coots showed high levels of radioactivity that had not been seen in decades.

The Chernobyl cooling pond is also a significant source for contamination of groundwater that then flows into the Pripyat River, but cleaning up the pond is practicably impossible. The costs, in terms of finances and occupational radiation doses, would be too high and there is no place to put the radioactive waste that would be recovered.

So, what to do with a cooling pond that no one needs and costs money to maintain? One option is to keep replenishing the pond with water so that it doesn't dry out. Another is to lower the water levels gradually and cover the exposed sediments with plants that will help prevent erosion.

Sediments aren't the only problem. The cooling pond contains about 100,000 tons of radioactive organisms that will eventually die and decay if it is allowed to dry out. That prospect poses not only sanitary problems but also radiological issues since the radionuclides will get out of the aquatic ecosystem where the water keeps them relatively isolated from human populations.

For example, much of the cooling pond's bottom is covered with zebra mussels. Named for the striped pattern on their shells, zebra mussels originated in Eastern Europe but expanded westward in the eighteenth and nineteenth centuries. Discovered in Canada in 1988, they have become a serious aquatic pest in North America, disrupting ecosystems and clogging water supply pipes of hydroelectric and nuclear power plants with their extremely high densities, which can reach as much as two to three kilograms per square meter.

Although each individual mussel is small—less than an inch in length—collectively they can play a decisive role in radionuclide cycles. Together, they filter six percent of the cooling pond's water each day. Filter feeders that tend to the bottom of the cooling pond, they are in close proximity to the radioactive sediments and filter them whenever they get stirred up. Also, since they are near the bottom of the food chain, they pass on radionuclides to the fish that eat them. Even in 2002, a kilogram of zebra mussels in the zone's most contaminated waterways contained as much as 27,000 becquerels of cesium and 62,000 becquerels of strontium. Strontium levels are especially high because about half of a normal mollusk's shell is made of calcium which, in Chernobyl mollusks, gets replaced by radiostrontium. Moreover, dead mollusk shells continue to adsorb cesium and strontium from water. Radionuclide levels in dead shells can exceed those in live ones. Fish species such as roach whose diet consists almost exclusively of zebra mussels are especially radioactive.

Zebra mussels in the cooling pond are not leaders in radioactivity levels. But there are simply so many of them that their total population is estimated to contain about 400 billion becquerels of strontium, or 11 curies.

The problem of exposing radioactive bottom sediments is not limited to the cooling pond. The flow of radionuclides from the Pripyat River into the Dnieper has broader implications. Were the Dnieper still a natural river, those radionuclides would have largely washed out into the Black Sea. And some, in fact, did. Indeed, a research expedition to drill the Black Sea bottom for radionuclides ended up discovering that around 5000 B.C.E., the Black Sea, which was then a freshwater lake, was suddenly and catastrophically flooded by saltwater from the Mediterranean. Some scientists speculate that its transformation from a relatively shallow lake into a deep and brackish sea may be the real event behind the biblical flood story. But in Soviet times the Dnieper River was transformed into something called the Dnieper Cascade—a series of six giant reservoirs connected by small sections of the natural river. To flood the 350 square miles needed to create the Kiev reservoir, dozens of villages southwest of the town of Chornobyl had to be evacuated from the banks of a 50-mile stretch of the Dnieper River. One of those villages was Teremtsi, which is an evacuated village in the zone.

The reason it was in the zone instead of being under water was that it was right on the border of the area to be drowned. So when the villagers petitioned Soviet authorities to let them stay, they received a positive response. After the reservoir was filled and fishermen could see the bell towers and domes of drowned churches in its depths, some Teremtsi residents returned—until they were evacuated again in the wake of Chernobyl.

As with many Soviet schemes to change nature, the Dnieper Cascade is largely an environmental disaster. The reservoirs are shallow, too warm, and plagued by algal blooms, and periodically there have been calls to drain them. But even when such proposals are seriously entertained—and they usually aren't—the bottom sediments of all the reservoirs are contaminated. Some predatory fish in the Kiev reservoir still exceed permissible radioactivity levels. But if the reservoirs are allowed to dry out without some kind of countermeasures to prevent their erosion, their radioactive sediments will eventually blow away in the wind.

RADIOACTIVE LAKES

Driving west on the cooling pond's dike, we passed about a dozen cam-ouflaged CheVE workers, and Sikora asked Leonid to stop the car so that he could check on them. He explained that they had put up a pumping station to drain a pond that formed when a protective dike was built on the right bank of the Pripyat River.

"The pond formed on a very contaminated area and when the water levels get high, they flood the nuclear station's basements. So, we set this emergency pump to pipe the excess water into the cooling pond. Eventually, we'll set up a permanent pumping station," explained Sikora, lighting a filterless cigarette before heading off to give some epithet-laced orders to the crew.

Given the cooling pond's uncertain future, it was not clear where exactly the permanent pumps would be placed or where they would drain the water. But the temporary pumping station was on a concrete platform built on an old canal. It ran parallel to the cooling pond's dike to drain the water that filters through the sand. It was only partly suc-cessful. In high-water years, most of the radionuclides that the Pripyat carries out annually are from flooding the contaminated floodplains and the polder. But in low-water years, radionuclides are contributed by other sources such as groundwater and polder waters that filter through the dikes. A quarter of these "low-water" radionuclides filter in from the cooling pond.

Ducks swam amid the tall cattails, with their familiar brown pok-ers of seeds waiting to burst in the spring, and some of the pump op-erators amused themselves by making duck calls and chuckling in response to their quacks.

One side of the canal was ringed with yellow flag iris plants, their seed capsules looking like small green bananas ready to split and re-lease their brown seeds. A young weeping willow was littering yellow leaves around them. But I was more interested in the water, which looked from a distance as though it was covered with pond scum. The green floating gook that is made up of blue-green algae, diatoms, and filamentous algae concentrates radionuclides. In the spring of 1987, some samples of pond scum in the cooling pond contained tens of millions of becquerels of cesium-137 per kilogram! By 1999, cesium levels had decreased significantly, but they still reached 200,000 becquerels at a time when most plants registered only a few tens of thousands.

When I clambered carefully down the uneven gravel that bordered the canal, instead of pond scum I saw what looked like a mat of green lima beans floating on the water. It was duckweed, one of the smallest and simplest of the flowering plants and a frequent resident of ponds, marshes, and quiet streams. The individual duckweed plants hang together in masses and often cover the surface of ponds.

My dosimeter rapidly squeaked a reading of about 200 microroentgens per hour when I crouched on the gravel by the canal's edge for a closer look. This was quite high, given that the canal was surrounded by fresh cement and gravel. Although I was tempted to try to pick out some duckweed for a closer look, the rocks were slippery. I was game for taking some risks for a good story, but falling into such a contaminated canal was not one of them. A damsel fly buzzed lazily over the water's surface, but there were few other insects given the early autumn temperatures.

Though each duckweed plant consists of a free-floating, oval frond less than three-eighths of an inch, the plants often appear to be in clusters of two to five. The clusters are actually duckweed reproducing asexually. Each duckweed plant is a single individual that produces buds from which new duckweed plants grow. For a time the offspring stay attached to the mom frond, forming temporary frond families.

Radiation seems to damage the plants' spatial orientation in frond formation. Like the twisted pine bushes that can be used to identify radioactive waste dumps, twisted duckweeds are signs of high radioactivity levels in water. Duckweed is either right- or left-sided, depending on the direction in which its daughter fronds grow. Some studies show that radiation inverts the "sidedness." Fronds in right-sided plants become left-sided and vice versa. The higher the radioactivity levels, the greater the percentage of inverted forms. Unlike twisted pine trees, however, these are forms that only scientists can see. If the duckweeds in the canal were inverted, I couldn't tell by looking at them.

Back in the van, we drove down a deep sand trail that passed through colorful autumn woods towards the Pripyat River, but at some point the going got too rough for the van and we climbed out to walk across what looked like a wide beach but was actually an alluvial dike built to prevent spring flooding of a highly radioactive patch on the right bank of the Pripyat River. Unlike the left-bank flood plain, which raised alarms in 1991, no one had worried too much about the right bank for much of the postdisaster period because it was at a relatively high elevation not vulnerable to flooding. But 1999 was a record year for spring flooding—the third-highest water levels in more than a century. The walls of CheVE's offices in Chornobyl are decorated with aerial photographs depicting complete inundation of the contaminated right bank.

A short distance away, a lake mirrored the sky as I followed Sikora through deep and loose sand that displayed fresh moose prints. The alluvial sand, pumped up from uncontaminated layers of the Pripyat River bottom, was scattered here and there with fragments of shell from freshwater mollusks such as zebra mussels and snails. With a diet that consists of detritus—bits of matter from decomposing organisms—the snails had very high radioactivity levels in their muscle in the first two years after the disaster because the radionuclides stuck to the surfaces of things in the water, including detritus. And since very tiny things like detritus have higher surface areas compared to large ones, detritus-eating organisms accumulated a lot of radioactivity. The levels in the snails' muscles went down in subsequent years when nearly all of the radionuclides washed off and sank deep into the bottom sediments. But like zebra mussels, the snails continued accumulating strontium-90 in their shells.

We were on the southeastern side of the highly contaminated patch, which did not itself have an official name. "Right-bank floodplain" was the only name I ever heard. Like the Red Forest and the polder, it suffered the heaviest fallout from the initial explosion and was colored brown on the radiation maps. Each of its five square kilometers was contaminated with as much as 1,600 curies of cesium and 450 curies of strontium.

On the edge of the lake, with the Sarcophagus looming over the horizon, Sikora showed me two shallow impressions that Chernobyl ecologists had experimentally planted with willow to see how it would prevent erosion if higher-ups in Kiev ever made a decision to drain the cooling pond. Grasses and a thick layer of thatch had also appeared.

"It keeps the wind from raising the radionuclides," Sikora observed. "But look at all that dry grass. Heaven forbid there's a fire."

Cattails, rushes, and wind-burned reeds grew thickly around the edges of Lake Azbuchyn, which ranks up there with Lake Hlyboke in the polder as one of the zone's most contaminated bodies of water, although the ranking depends on which radionuclide is measured. Plants in Lake Hlyboke have the highest cesium-137 levels in the zone, running up to 36,000 becquerels per kilogram in 2003. But plants in Lake Azbuchyn have the highest strontium levels, with an average of 4,300 becquerels per kilo and a maximum of 24,000.

Actually, the average radioactivity levels are not very informative. If I collected a bunch of rushes from the same place on the edge of Lake Azbuchyn, there would be huge differences in their radioactivity. Indeed, the differences could be as high as 140 times because of the patchy contamination of the bottom sediments. Nevertheless, higher plants are more convenient for monitoring aquatic radioactivity than, say, fish. Fish move around, and therefore sampling always involves a certain randomness, while plants stay put.

As a rule, emergent plants such as sedges, reeds, and rushes—which are rooted in the underwater sediments where radionuclides are concentrated but emerge from water into the air—have the highest levels of cesium. The lowest levels are in plants that float on the surface of the water. But this is not true of strontium. Its highest levels are found in pondweed, some species of which have floating leaves that intensively absorb calcium—or its Chernobyl substitute, strontium—on their surfaces during photosynthesis. This makes them promising plants for radiation monitoring.

Unlike aquatic animals, which continue to accumulate radiostrontium in the course of their lifetime—and in Lake Azbuchyn the average strontium concentrations in a kilo of zebra mussels exceeded 50,000 becquerels in 2002—the aquatic plants were beginning to lose theirs as the growing season drew to a close. This could be because parts of the plants were dying and the oldest parts, with the highest accumulations of strontium, were falling to the bottom of the lake. The cooling waters were slowing biological processes, making strontium less soluble and affecting the plants' ability to absorb it. The shortening days were also decreasing photosynthesis, which in turn was slowing the plants' accumulation of nutrients and of the radionuclides that mimic them. The real reason could be all of the above or none. Because of budgetary shortfalls, there is barely enough money to monitor the radiological situation and almost none to study it.

My dosimeter displayed 1,000 microroentgens (or one milliroentgen) an hour when I followed Sikora up a moss-covered concrete staircase that climbed a steep, grassy incline. It was a hot spot indeed, although I had no idea of what he planned to show me until we reached the top and I saw that it was the railroad track of the Slavutich-Chernobyl line. The elevation provided a panoramic view of the Pripyat River's floodplain. The track must have been decontaminated because the dosimeter calmed its excited squealing and dropped to a very normal 13 microroentgens.

Sikora pointed north of the tracks at a wide expanse of sand that peeked in and out of the vegetation on the Pripyat's riverbanks. He explained that the sand was a new, two-and-a-half mile dike built to protect the right-bank floodplain. It connected with the cooling pond's dike that was beyond the forest south of the tracks, enclosing the patch of contamination. Although the need to protect the patch was recognized in 1999, the dike wasn't actually completed until 2004 because there was no financing.

Another problem is that the Pripyat is an active, young river. After the last Ice Age, glacial meltwaters turned Polissia into a huge freshwater lake known as the Polissia Sea that was eventually transformed into wetlands and the Pripyat River basin, with its numerous tributaries, enormous floodplain, complex and diverse landscape, and large number of floodplain lakes, bays, and tiny streams.

The river continues to actively dig its channel with the friction of flowing water on its banks and bed, but parts of those banks are contaminated with up to 100 curies of strontium per square kilometer and more. So, channel digging and the attendant erosion spill contamination into the river. A 4.7-kilometer section of the river between the railroad bridge and the village of Kosharovka had to be reinforced to prevent this.

"And that's Lake Daleke," Sikora said, pointing south of the tracks at another lake that was even closer to the plant than Lake Azbuchyn, though it was marginally less contaminated. In nearly all the radiological monitoring of cesium and strontium levels in the zone's waterways, Lakes Azbuchyn and Hlyboke vie for first place, while Daleke is usually a close third.

Some strangely sexy things seem to happen in those lakes, too. In what may be an effort to resist the effects of radiation, aquatic worms that are capable of both sexual and asexual reproduction are more likely to have sex in Chernobyl than control worms outside the zone. Asexual reproduction means that each worm's offspring is stuck with whatever genes its parent had, while sexual reproduction causes genetic mixing that allows natural selection to promote genes and gene combinations that increase resistance.

The lakes reflected fall foliage under the brilliant blue of the October sky, and I thought it hard to believe that something so beautiful could be so deadly. Yet, as we walked back to the CheVE van, I wondered if it was correct to call them deadly at all—although "lethal," "deadly," and similar adjectives are often used to refer to radioactivity, whatever the amount or the source. But my body is radioactive, as is yours and that of every single person on the planet. And while I may, on occasion, get very angry, I am not deadly.

Yes, the lakes were radioactive, and no, I would not go swimming in them or drink their waters and I didn't even want to be near them for very long. But they were not radioactive enough to kill with any certainty of time or place. The plants and animals in their waters were very much alive and, given the lack of human disturbance, might be even more plentiful than if the disaster had never happened. I was reminded of the restored—and very radioactive—peat mires in Belarus that had become beautiful and haunting magnets for aquatic birds.

As we drove past the decommissioned nuclear plant that had caused so much human heartache, I had mixed feelings about thinking something positive could have come of the disaster. But then I wasn't sure if a radioactive nature preserve was a good thing or a bad thing. I wasn't even sure if it could be called "natural" if it was radioactive, not because of the slight radioactivity with which creation endowed our planet and which is a part of us all, but because of the great (and truly deadly) amounts of the stuff produced by human effort and error. I was certain, however, that nature was refusing to abide by the absolute definitions we try to foist on it. Endangered species rebound in war zones, while grizzly bears struggle to maintain their numbers in protected areas such as Yellowstone Park. Chernobyl wildlife was thriving.

The van bounced and creaked over a paved, but rough, road that led from Chernobyl to a peninsula on the Pripyat that formed the *Yaniv Zatok*, or Yaniv Bay, named after the buried village of Yaniv just outside the nuclear station's grounds. In nearly all the literature I had read about Chernobyl's impact on aquatic systems, it was always mentioned as a significant danger. But like the articles about the left-bank polder, little of the literature about Yaniv Bay described the problem in a way that was comprehensible to someone who wasn't already an expert. All I knew was that it was a comma-shaped arm of the river, directly across from the polder, and stained the same shade of brown on the contamination maps.

The landscape on either side of the road was littered with sandy humps and bumps topped with triangular orange trefoil signs and Cyrillic letters that transliterate as PTLRW—the points for the temporary location of radioactive waste that were Valery Antropov's biggest headaches in Chapter 1. Antropov had told me then that they used to think there were 800 of them, though in fact they had lost count and many had become flattened and thus invisible with time. Wherever they all were, they seemed to be everywhere alongside the road to Yaniv Bay. After more than a dozen trips to Chernobyl, I had never seen so many leaky radioactive waste dumps in one place.

Three tall cranes that used to unload ships in the Pripyat town port were silhouetted behind the deep crimson of some maple trees. A huge flock of migrating cormorants dotted their rusting and radioactive cables like musical notes. I asked Leonid to stop so that I could take a closer look with my binoculars and also measured out an hourly reading of 200 microroentgens on my dosimeter.

We went up a steep incline of deep sand, and Leonid parked the van near a small blue building with a gabled roof and white trim. Its cheerful colors looked incongruous on the desolate sands surrounding us when Sikora led me to the entrance and unlocked the metal door that announced the structure was paid for by the United Nations Development Agency in 2002. There was nothing inside except a simple little room, about 10 feet square, containing a wooden table, a chair, a broom, and a gauge on the wall.

Sikora explained that the system also included a metal plate attached to the riverbed and a sensor that regularly sends the plate an acoustic signal. The sensor determined the water level on the basis of how long it took the signal to bounce back.

"We send someone here once a week to download the data into a laptop," said Sikora, leading me out and locking the door.

We crossed a broad expanse of sand that narrowed into a dike imprinted deeply with tractor treads. It was built right after the disaster, together with more than 130 other hydrological systems intended to prevent flooding and radioactive runoff. But they were built at a time when no one in the world had experience in cleaning up such a disaster and nearly all of the systems quickly silted up, leading to even more flooding. Almost all of them were taken apart and replaced with new ones.

One of the few to stand the test of time was Dike No. 3—the one we were walking on. After walking for about five minutes through a forest of young (and normal-looking) pines, we came to the place where the dike cut off Yaniv Bay from the Pripyat River. Sikora explained that the purpose was to prevent spring floods from raising waters levels in the bay because the lands around it were littered with all the PTLRW we had passed on the way there. When water levels in the bay are high, they flood the radioactive waste, flushing radionuclides into the water. The dike was supposed to prevent them from ever getting that high.

"Why did they bury radioactive waste here, by the water?" I asked Sikora, although I knew what the answer would be.

He shrugged in response. "No one was thinking about the ramifications then."

Although Dike No. 3 had served its purpose quite well, it didn't completely prevent water from filtering through the sand, which was why, when we arrived, there was a confusing bustle of CheVE activity involving trucks, earth movers, and small mountains of alluvial sand. Diesel fumes clung to the air when I followed Sikora to a freshly dug trench containing a big black pipe that lay perpendicular to the dike's length.

"This pipe will eventually link the river and the bay," he told me as I followed him towards the bay across a temporary path of flat blocks covered with mosaic tile, as though they had come from a swimming pool. He showed me a large box with mysterious innards that he said would automatically drain water out of the bay whenever it exceeded a certain level.

Ringed with scarlet maples, yellow birches, and the beiges of beech trees, the bay was lovely in its autumn reflections. But through my binoculars, I could see some of the little triangular trefoil signs that dotted the woods around its shores. They had some elevation and weren't right on the water.

For a time, at least, it seems that they won't get flooded. But they do pose other dangers to the water.

ATOMIC AQUIFERS

Semykhody was a ghost village long before the Chernobyl reactor exploded. Although it is on all the topographical maps in a place that appears to be located right on top of the nuclear station, there is actually no evidence of its existence. The Chernobyl cultural expeditions that the Ukrainian Ministry of Emergencies sponsors to collect artifacts from abandoned villages have been looking for evidence of Semykhody's existence for years, without success.

But place names are far more enduring than the memories of those who bestowed them, and even the memories *about* those who bestowed them, which is why another bay of the Pripyat River is named for Semykhody. Located about a mile upriver from Yaniv Bay, Semykhody Bay is much smaller but about equally contaminated. Like Yaniv Bay, it is also cut off from the river by a sandy dike.

We had to drive around the town of Pripyat to get there, passing the nuclear plant and a construction site on the road where a deep excavation clearly showed the new layers of roadway that were put down after the disaster to cover the contaminated surfaces. Actually, it was very rare to see any kind of roadwork in the zone except for patched potholes here and there on the main roads in Chornobyl and around the nuclear station. Even pothole patching was inconsistent.

But Sikora explained that the excavation was not to fix the road. It was to lay pipes to a new decontamination facility for the nuclear workers who would eventually work on decommissioning the shut-down reactors and dealing with the radioactive mess inside the Sarcophagus.

We were heading for a place informally called the Sandy Plateau, which does not appear on the topographical maps but ended up looking exactly as its name would suggest. To create new land for the town of Pripyat's planned expansion, several acres of alluvial sand had been pumped out of the Pripyat River between Yaniv and Semykhody Bays. I forgot my dosimeter in the van when I trudged through the sand behind Sikora, so I didn't know what the radioactivity levels were, but the area was colored the darkest shade of red on the radiation maps, meaning that the disaster dumped around 500 curies of cesium there in 1986. There was actually almost nothing interesting to see on the Sandy Plateau, but the reason I went there was not because of what could be seen but because of what was invisible—or at least invisible to the unaided eye.

For about 15 years after the Chernobyl reactor exploded, the zone was considered a kind of protective radiation sink that kept radionuclides relatively fixed in the layers of soil and sediments where their harm to the general population could be minimized. But this changed in 2002 when nuclides, especially strontium-90, started appearing in the groundwater. One of the places where such contamination was detected was the Sandy Plateau.

Probably like many people, I have always imagined groundwater to resemble vast underground lakes or rivers that are all interconnected and that we dip into with buckets and pipes. So, when I first read that strontium had penetrated the water table—which is the top of the groundwater—I imagined these subterranean flows of radioactivity that would spread into drinking water far beyond the zone, including my own faucet in Kiev. In fact, the groundwater is in layers of permeable rocks with fractures and pores that hold water from precipitation like sponges. When those waters are easily transferred to wells and springs, the water-bearing rocks are called aquifers.

The zone has three primary aquifers, all created by events that occurred many millions of years ago and containing water that can be hundreds and sometimes thousands of years old. The town of Chornobyl gets its water from rocks deposited during the Eocene period around 45 million years ago. The nuclear station's water supply is from Cretaceous rocks deposited around 150 million years ago, during the age of dinosaurs. More than 18 years after the disaster, cesium and strontium levels in a cubic meter of these aquifers were less than 25 becquerels, which is so low as to be within the margin of error. But the quaternary rocks that were deposited in the last 2 million years and are therefore in the shallowest layers are a different matter. In some places, these quaternary layers are becoming contaminated.

Where exactly depends on the source of the radioactivity. There are two major sources in the zone and two ways that it can get into the groundwater. The "diffuse" source basically consists of the entire zone and all of the radionuclides that are now in virtually everything in the environment: the soil, plants, and animals. They contribute little to groundwater contamination. In nearly all places in the zone, strontium levels in a cubic meter of groundwater are usually double-digit becquerels and don't exceed a few hundred. Cesium levels are even lower.

The problem is from so-called point sources: the leaky nuclear waste dumps in the hundreds of PTLRW around the zone. Highly contaminated sediments in the radioactive lakes and bays such as Semykhody and Azbuchyn are also point sources, as are the Sarcophagus and the very grounds of the nuclear plant. Although the reactor debris that littered the grounds was bulldozed and the territory was covered with fresh asphalt and concrete during the cleanup, a good deal of radioactive material remains in the soil below, where drilling has revealed groundwater contamination.

After the turn of the third millennium, significant levels of radioactivity began appearing in quaternary groundwater deposits beneath all of these places. This is another reason Yaniv Bay is so dangerous, even if the new pumping station helps prevent flooding of the waste dumps. Rain trickling through the radioactive garbage washes radionuclides into the soil, and from there some travel into the groundwater. In 2000, test drills in Yaniv Bay found a maximum of 400,000 becquerels of strontium in a cubic meter of groundwater! Levels were lower in 2002—270,000—but this is still more than 100,000 times maximum permissible levels. Similar contamination levels are found in the groundwater beneath the Red Forest—where the buried trees that took the deadliest blow of radioactivity are gradually decomposing, releasing their vast stores of radionuclides into the soil. But Yaniv Bay is much more dangerous because of the Pripyat River's proximity.

No other point sources of groundwater contamination have such high densities of radioactivity. The maximum strontium level measured under the cooling pond, for example, was 4,800 becquerels in a cubic meter. Another point source of radioactivity is the official waste disposal site in Burakivka, which I visited in Chapter 1. The clay seals there seem to be working quite well, and contamination of the ground beneath them is within safety limits. Some scientists nevertheless caution that the network of drills at the site doesn't allow monitoring of groundwater under any of the trenches and there are no drills at all between some rows of trenches.

Most groundwater contamination remains localized, right under the point source. Yet although it is hard to generalize about groundwater, it does not always stay in one place. When it does move, it usually goes downwards because gravity pulls it towards the center of the Earth, but it can also move sideways. A particular aquifer's qualities—its thickness, depth, the type of rock, and a multitude of other factors that make up what are very complex systems—determine how and where the groundwater will move. In the zone, the groundwater's lateral flow is generally towards the three main rivers—the Pripyat, Uzh, and Braginka—all of which drain into the Kiev reservoir. But all of the point sources of groundwater contamination are in the Pripyat River basin.

How much of the strontium in that water actually gets into the river and thus into the water supplies down the Dnieper Cascade is uncertain. Ukrainian geologists predict that radionuclide migration under the point sources will intensify 20 to 30 years after the disaster, creating a multitude of groundwater plumes that will constantly pour contamination into the Pripyat. However, some think that there will not be a great risk outside the zone because the radionuclide resides in the groundwater for a long time during which some of it decays. What doesn't decay will dilute as it travels downriver.

When groundwater gets contaminated, however, it's not always clear why. For example, one drill in the Sandy Plateau, not far from the Pripyat River, found 100,000 becquerels of strontium in a cubic meter of groundwater that was 15 meters (50 feet) deep. At a depth of four meters (13 feet), the level was 130,000! Anomalously high radioactivity levels have also been detected on the riverbank between the Sandy Plateau and the railroad bridge about a mile downstream. Scientists aren't sure exactly why this should be so. Although it is highly contaminated, the Sandy Plateau is a diffuse rather than a point source of radioactivity and could not have been the origin of such high levels of radioactivity in groundwater.

One possible explanation I read was that the concentrated radionuclides may be from groundwater flowing to the river. But when I asked Sikora about this, he seemed dubious and dug into the sand with his heel. "There is an old sewage pipe underneath here that connects Yaniv and Semykhody Bays. It was laid before the accident when they were preparing to expand Pripyat. But it's on the bottom of the bays where all the radionuclides settled. It's probably leaking."

It was hardly the first time that I had read something in a scientific journal, taken copious notes, and then traveled to Chernobyl to see it firsthand only to find out that what I read might be completely wrong. Or it might not be. The scientists who wrote the articles might not know about the sewage pipe. Or Sikora might not know the scientists or the results of their research. When I told him of the groundwater contamination under the Sandy Plateau, it was news to him.

I do not mention this to doubt Sikora's word. Not once during any of my Chernobyl travels did I ever feel that he or anyone was deliberately lying or misleading me. Rather, I mention it to point out how much about Chernobyl remains unknown and disputed.

And perhaps one of the greatest mysteries is the disaster's impact on people.

7

Homo chernobylus

Here live the owners
And our kind will not be moved

—Signs on zone homes

For centuries, Polissia was perceived as poor, illiterate, and backward—at least in the view of the government officials occasionally sent there by distant capitals. It was a picturesque periphery that was never at the center of much of anything and produced nothing and no one of note. Except for Jews and Germans—whose ethnic identities were distinct—the region was peopled largely by Slavs who were rather vague about whether they were Poles, Ukrainians, or Belarusians. Usually they just called themselves *tuteshni*, or "locals," even after Soviet bureaucrats assigned them with standardized—and, in many cases, largely arbitrary—ethnicities in the 1920s.

Getting assigned an ethnicity was, nevertheless, benign compared to what came afterwards. In 1932-1933, the Soviet dictator Josef Stalin deported the so-called kulaks, who were basically any peasants who were not poverty stricken. Forced collectivization herded those who were left into state farms and an artificial famine, which killed up to 10 million in Ukraine, broke the resistance of those who refused. A few years later, ethnic Poles were declared enemy agents of interwar Poland and deported to the Soviet interior as were Germans accused en masse of spying for the Nazis. Then the Holocaust wiped out much of the Jewish population of the German-occupied USSR, including the Jews of Chornobyl—home to a rabbinical dynasty that today attracts occasional Hasidic tourists disappointed to find that the Soviets long ago paved over the old Jewish cemetery with a parking lot.

Despite the repression and upheavals, however, most rural people had lived in Polissia for generations. After Chernobyl, many were reluctant to leave behind their land and their livestock, fowl, and pets because of some invisible radiation that few of them understood. There were many incidents of people, especially the elderly, hiding in cellars, haylofts, and forests to avoid evacuation. Two old women spent a month in Pripyat after its evacuation, living on canned food and bottled water before being discovered and resettled.

The zone's barbed wire perimeter didn't prevent the rural folk from sneaking back to their old homes. Told that they would be gone for only three months, some began returning in the summer and fall of 1986 to tend their involuntarily abandoned cows, pigs, and ducks. And they kept coming even after all the contaminated animals were removed or slaughtered. After some attempts to make them leave, zone authorities decided to let people over the age of 50 stay—but only in the Ukrainian part of the zone. Their numbers have been declining steadily.

Called *samosels*, which means "squatters" or, literally, self-settlers, they are the zone's only permanent residents.

But they are far from its only people.

VIRTUAL VAGRANTS

The bridge from the town of Chornobyl crosses the Pripyat River a few miles south of the defunct nuclear plant's cooling pond, where the river channel branches out and braids around islands and swamps on its way to the archipelago of tiny islands at the delta. A ribbon of brilliant sapphire reflecting the sky in its main channel, the Pripyat seemed to fragment into shards of mirror that gleamed amid the textured emerald of the swamps.

I was driving a carload of visitors on a warm June day, though it wasn't exactly clear who was tagging along with whom. Rimma Kyselytsia of Chernobylinterinform was our guide, but I think she learned as much from her guests on that trip as we did from her. Wayne Scott was an American birding enthusiast and naturalist. Kate Brown was an assistant professor at the University of Maryland–Baltimore and the author of *A Biography of No Place*, an affecting history of the Polissia region. I wanted to visit with some *samosels* and my plan was to take the road I had not yet traveled on the left bank—namely, the one that led southeast to the Kiev Sea.

While we waited in the car for the guard at the Paryshiv-2 checkpoint to open the gate, one of the stray dogs that seem to attach themselves to all Chernobyl checkpoints waggled submissively for affection. It was a skinny bitch whose hanging teats indicated a litter of puppies not far away.

Wayne pointed out some black storks circling over the trees in the distance. Although they are far more common on the Belarus side of the zone, black storks have also been appearing in Ukraine. Unlike their white cousins, black storks like the wild country of forests and swamps where human intrusions are rare. Before Chernobyl, there were no black storks in the area at all. They appeared after the evacuation, and their Ukrainian population had grown to about 40 in 2000.

For someone like me, who enjoys birding but can only identify most species if the bird sits still, in good light, for several minutes— and that's with the help of a handbook—Wayne's ability to do so quickly from a distant silhouette was a useful skill in a traveling companion.

The village of Paryshiv was about half a mile from the checkpoint, on a pale yellow patch on the radiation maps. Twenty-two people lived there—making it moderately populated as zone places go. Aside from Savichi in Belarus, a sizable settlement of 120 people that includes 12 children, and the town of Chornobyl, where up to 150 people live depending on the season, zone village populations range in size from 2 to 50.

Paryshiv looked largely abandoned when we stopped there. The old cottages visible from the road were rotting. Roofs had collapsed, breaking floorboards, and vegetation had begun to blanket the sharp edges. Some buildings could barely be seen through the thick overgrowth. Kate clearly wanted to explore, but whenever she went too close to some shaky-looking structure, Rimma pleaded with her to stop, with me chiming in my own concerns.

"I wasn't going to go inside," Kate tried to assure us. But she went a lot closer than I would have. The ground in the abandoned villages is deceptively solid, concealing overgrown manholes, wells, rusty nails, and other treacherous things that can easily do you bodily harm. I was picking up some extra lifetime radiation doses for my story. I had no intention of risking broken limbs or tetanus to boot.

Kate was apparently an intrepid, Indiana Jones sort of professor. But she deferred to our concerns and asked if we could visit the village graveyard.

Polissian graveyards are not always easy to find since they are usually tucked in the woods. Rimma suggested we go four miles to the next village, called Ladyzhychi, where the cemetery was right off the road.

"I read about some cemetery that's full of Smirnov surnames," said Kate on the drive there.

That sounded familiar to me. "That's not from that Chernobyl biker chick Web site, is it?" I asked. In the spring of 2004, a Kiev woman who claimed to ride around the zone alone on her Kawasaki motorcycle created a Web site that captivated the chattering classes on the Internet.

"Yes, it is," said Kate, prompting Rimma to interject vehemently.

"That story is a lie! She never rode a bike here. She came in one of *our* cars and with her husband! She just carried a motorcycle helmet around and he took all the pictures."

It was such a strange thing to lie about. But in the spring and summer of 2004, a number of marketing and entertainment enterprises decided to exploit Chernobyl's inherent spookiness. For example, opening scenes from *Necropolis,* the fourth installment of the *Return of the Living Dead* movies, were filmed there—despite the protests of the zone administration, which didn't like the associations it provoked. People I talked to suspected that the filmmakers paid some hefty bribes in Kiev.

In general, the number of zone visitors has been increasing steadily, pretty much in direct proportion to falling radiation levels. In 1995 about 900 people visited in the entire year. In 2004 there were 2,600 visitors, and these were just the delegations of scientists, officials, and journalists. If you include evacuees and their relatives, 30,000 people visit the Ukrainian zone annually.

There are no figures for the Belarusian reserve, which doesn't keep track of visitors.

The number of pure tourists has been growing too, and a 2002 United Nations report suggested encouraging "ecological tourism" as a way of bringing money to the region. Concluding that after 15 years,

"Chernobyl"—and all that the word entails—was no longer an emergency requiring urgent measures to help victims, but had entered into a 10-year recovery phase during which the victims should be helped to feel empowered and in control of their lives, the report urged the governments of the affected areas to help people learn to live safely—and even profitably—in radioactive environments.

This seeming contradiction provoked a lot of derisive commentary. But after nearly two decades of decay, environmental radioactivity has fallen to less than one percent of the total amount released. Since nearly all of this is in the soil rather than on surfaces, external radiation doses are no longer very high, even in regions that were highly contaminated in 1986.

Ladyzhychi, like Paryshiv, was on a yellow patch. After we pulled up in a shady spot outside the graveyard, Wayne went off to look for birds while the rest of us explored the cemetery.

Judging by the tombstones, the village had been populated entirely by people whose surnames were Melnychenko. I stopped by a handsome black granite tombstone marking the grave of Mykola Melnychenko, who died at the age of 69 in 1997. He was buried next to his wife Katerina, who died two years later. It wasn't clear if they had returned to live in the zone or were just buried there. Many evacuees asked to be buried in their native villages.

Coincidentally, Mykola Melnychenko was also the name of the ex-bodyguard who taped Ukraine's President Leonid Kuchma vulgarly ordering his underlings to intimidate a muckraking journalist whose headless body was found in the fall of 2000. The worst political scandal in Ukraine's history, "Tapegate" exposed all the ugliness of a corrupt regime whose misuse of power and public funds was a big reason there was no money for Chernobyl works and research.

An old, rotting oak had what looked like a log strapped to one of its branches. Actually, it was a wild beehive that had been abandoned by its bees. Someone had sawed above and below the hive in its original tree home and then brought the hive-in-a-log—called a *bort*—to a more convenient location. Cemeteries were a popular place to put the wild hives because they were accessible yet quiet.

Wild beekeeping, called *bortnytstvo* in Ukrainian, is the most archaic of apiary arts. In predisaster Polissia, the methods had changed little since the tenth century, when one of the region's main exports was honey shipped down the Dnieper River to Byzantium. In general, Polissia was a very archaic place well into Soviet times.

Aside from about a hundred graves, the cemetery contained several empty *bort'*, what seemed to be about a zillion mosquitoes, and a pile of rotting baskets and tools. Kate picked up a wooden rake and I switched my dosimeter to measure beta radiation. It beeped out a digital display of 150. This meant that a square centimeter of the rake was shooting out 150 beta particles per minute. The maximum safe limit was 20. The rake wasn't exactly "hot." In Chapter 3, for example, Igor Chizewsky had shown me animal skulls emitting several thousand beta particles. But it was "warm" enough and Kate quickly put it down.

Our explorations were interrupted when Wayne showed up in what appeared to be police custody. A militiaman had found him wandering around alone in the village and had tried to ask him what he was doing there. Wayne spoke no Ukrainian or Russian, but he had somehow persuaded the policeman to follow him to the cemetery so that we could explain his presence.

Radioactive birdwatching was clearly a new idea of fun for that particular officer of the law (and not only him, I'm sure), but he accepted it cheerfully enough when Rimma showed him our program which set forth, in black and white, that one of the activities we had permission for was "birdwatching in various zone locations."

After he left, Rimma said: "There are all these rumors about the zone being a refuge for criminals and fugitives. But look, Wayne couldn't even wander alone for more than 15 minutes without getting caught." But Wayne also lacked the language and cultural skills to bribe the militiaman had he been so inclined.

Nevertheless, the Ukrainian tabloid press frequently made sensational—and unconfirmed—reports about the zone being an international transit point for illegal trafficking, a place for growing radioactive marijuana and opium poppies, a dumping ground for the corpses of murder victims, and God only knows what other illicit activities. In March of 2004, zone militia mounted a special operation to search a dozen ghost villages for a fugitive who had been wanted for months. He was finally apprehended in Tovstiy Lis, an abandoned village bordering the highly contaminated western arrow of debris from the initial explosion.

The fact that there are no border patrols on the Belarus-Ukrainian border also makes it a tempting place for illegal immigrants. In 2001, nine people from Afghanistan were detained illegally while trying to cross the border in Ukraine; they planned to settle in Kiev.

Because there aren't enough people to police them, large swathes of land are essentially lawless. This doesn't mean that they are crawling with criminals; it just means that no one necessarily knows what's happening in much of the zone at any given time.

Although zone officials maintain that the crime rate is lower than that of similarly sized territories in Ukraine, looting has been a problem since the evacuation, and little of value is left in the zone's villages and towns. Just six months after the disaster—when things left behind were very radioactive—two men were arrested for robbing Pripyat apartments of televisions, cameras, towels, cigarettes, candy, and food that they tried to sell without getting it checked for contamination. Scrap metal scavengers together with poachers—especially fisherman, whose fishing poles make less noise than hunting rifles—are the greatest crime problem.

Another form of illegal activity is simply being in the zone without permission. During a routine two-day inspection of the Ukrainian zone in the autumn of 2004, militia found more than 40 illegals, including two poachers who shot a moose. But trespassing alone carries no penalties, not even a fine. Violators merely get tossed out. In 2004 a man tried to avoid his alimony payments by hiding in Pripyat. Instead of expelling him, zone officials decided it would better to give him a job so that he could make his alimony payments.

Undoubtedly, other unauthorized people have also lived in the zone's abandoned villages and towns, though no one knows exactly how many or when—unless they are caught or die there and their bodies are found, like the anonymous old man whose drowned body was found in the Pripyat River, when it flowed through the village of Otashiv in the summer of 2003. He is buried in the Chornobyl town graveyard, together with about 15 other anonymous bodies found since 1986.

The zone's vagrants are a bit like virtual particles—the highly energetic subatomic particles that come in and out of existence but only for the tiniest increments of time. They can't be detected and, in a sense, escape reality's notice. But it is easier for zone vagrants to elude detection in the summer. In the winter, it's impossible to survive without fire, and the smoke can be seen from a distance.

SQUATTING AT THE DELTA

After leaving the cemetery, I checked my map to find the village of Teremtsi perched on a peninsula right on the Pripyat delta. It was the village that escaped drowning when the reservoir was filled in 1968. Beyond it there was nothing but the shallow waters of the artificial Kiev Sea. It was literally the end of the road and everyone agreed that it sounded like a good place to visit.

Teremtsi boasted 37 inhabitants in 2004, placing it among the zone's most populous villages. The *samosel* population has been dropping steadily since the postevacuation high of 1,210 people in 1987. In 1995 there were 820 people. On an average day at the start of the third millennium, about 300 people permanently live in the zone. Though some have moved away, most have died.

Few of those who remain want to go anywhere. In 1999, when there were more than 600 *samosels,* nearly 100 of them wanted to relocate—but when their new accommodations were built, most refused to move. When questioned again five years later, only 4 out of nearly 400 people said they wanted to be relocated outside the zone. From a different perspective, in a 1996 poll the vast majority of evacuees over the age of 50 who had remained in their new locations wanted to return to their homes in the 30-kilometer zone.

Although they were told that it was not a safe place to live, the *samosels* just kept coming back—sometimes avoiding checkpoints altogether. Zone officials did consider relocating them forcibly, but the prosecutors—with the support of Ukraine's Supreme Court—refused on the grounds that they had a constitutional right to live where they wished. So they exist in a shadowy state of semilegality.

Their villages are not supposed to get any public works—because officially, no one lives in them. But because people do, in fact, live in them, the administration can't just leave them on their own completely. So it maintains the roads that lead to their villages, delivers pensions and mail, ensures that there is at least one phone in case of emergencies, and checks the food they grow for radiation levels.

Once a month, residents can take a bus to Ivankiv to buy a piglet at the market or some new shoes. Traveling medical brigades give them checkups, while they get treatment for more serious illnesses at the special Chernobyl clinics and hospitals in Kiev. On major religious holidays, the administration buses them to the zone's sole working church. In a way, Chernobyl pensioners get better care than many of their counterparts outside the zone.

In contrast to Ukraine looking through its bureaucratic fingers, Belarus completely banned unauthorized habitation. But it did allow legal settlement in two villages on the radiological reserve's borders. Tulgovichi, the very radioactive place I visited in Chapter 3, was one of them. The other was Savichi. In Belarus there are no *samosels* because squatting is not tolerated the way it is in Ukraine.

The road ran parallel to the delta and its watery islands and swamps. When Wayne spotted a white egret, I stopped the car so that he could set up his telescope on the shoulder for a better look. The perfect bird for my birdwatching skills, the egret obligingly stood motionless amid the reeds for the five minutes we all took turns peering into the telescope.

Back in the car and driving east on a gutted asphalt road, past marshes and thick forests of old-growth trees, Kate pointed out an old wood cabin deep in the swamps. It was a *khutir*, or independent homestead, which had been quite common in Polissian parts until well into the Soviet era. Spaced widely apart and often accessible only by wooden bridges or pathways known only to the swamp dwellers, the *khutir* helped Polissians avoid forced collectivization longer than other regions of Ukraine. If they didn't want to give up their livestock and crops to the Soviet tax collectors, they just hid in the forests and swamps. Their knowledge of the fruits of the forest also helped them survive the artificial famine.

"It looks like there should be a footbridge leading to it," said Kate.

I turned onto a forest path that looked as though it led there. But it was overgrown and the bushes threatened to scratch the car, so I backed out onto the main road, thinking that a UAZ jeep would have been handy for the journey. It would have been even handier after the paved road ended and was replaced by a trail of deep, loose sand gouged into gullies and trenches by passing vehicles—like an exhaust-belching truck we saw hauling deadwood. It scraped the bottom of the car as we bounced and bumped for what seemed like forever.

"Just don't stop," Wayne advised. "Keep up the momentum or we'll get stuck in the sand."

It was too late to turn back and probably impossible without getting bogged down. So I pushed ahead, clutching the steering wheel with sweaty hands while trying to ignore the alarming clinking and clunking from below that seemed to signify that I was leaving behind a trail of car parts in the sand.

"Now, try to tell me that that Internet biker chick could ride a motorcycle here," I joked after we finally emerged back onto the road. It was potholed, cracked, and buckled. But it was paved and we could finally move faster, crossing paths along the way with a snake, a hedgehog, and a roebuck that gleamed like copper in the sun. The meadows, forests, and swamps were rich in marsh harriers and lapwings, yellowhammers, and hoopoes that Wayne identified with a quick glance or a few notes of birdsong.

Soon we passed the sign for Teremtsi. Most of the village was a crumbling collection of vegetation-choked cottages, but a handful of houses displayed signs of habitation: freshly chopped wood, tidy yards, flocks of chickens, geese, and ducks. Motrona Ilyenok and her husband Oleksi lived in one immaculately kept household.

Oleksi smiled a shy welcome when we knocked on his gate and walked into the courtyard. All of the walkways were lined with flagstones and bordered by a wattle-and-daub fence with ceramic pitchers draining on the posts. It was a classic image from Ukrainian rural culture whose kitsch versions decorate the "retro" restaurants of traditional ethnic cuisine that have become extremely popular in Kiev. Kitsch, however, was garish, while all the handiwork in the Ilyenok's household was in a palette of gentle pastels. It was like entering a beautiful but faded photograph.

Motrona invited us inside their spotless, two-room cottage. It was fully furnished and decorated with cross-stitch embroideries, icons, and framed photomontages of family and friends from years past. Their neighbor Maria Chala was visiting. Unlike the Ilyenoks, who lived in Teremtsi year-round, Chala stayed with her children in Kiev in the winter and went to the village in the summer time. Her Teremtsi home was more like a *dacha*, or country home.

The Chernobyl zone probably sounds like a bizarre place for a country home. But Teremtsi was actually in a clean area. My dosimeter displayed a microroentgen reading in the single digits—even lower than the streets of Kiev. In fact, Teremtsi was just outside the geometric borders of the 30-kilometer zone, although it was within the administrative borders of the Zone of Exclusion and Zone of Unconditional (Mandatory) Resettlement. The latter was a larger region that provided

what would be a perfectly circular border with a more jagged and meandering outline. On the radiation maps, Teremtsi was a light shade of
green. Less than two curies of cesium-137 sprinkled every square kilometer of it in 1986.

"We never should have been evacuated," said Motrona.

Radiologically, that may have been true. Yet maintaining utilities
and services for a single village at the end of a road that runs through a
radioactive zone makes little sense when all of its residents are poor
and public services must be subsidized.

The pull of the land may seem surprising, given its radioactivity.
But Ukrainian evacuees—especially those over the age of 50—suffered
much higher stress than the *samosels*. People who had lived their whole
lives in their own wooden cottages in the cool, lush forests of Polissia
suddenly found themselves crowded together with several other families in shoddy, hastily thrown up housing in muddy villages on the
steppe or on the outskirts of towns and cities. Often they didn't feel
very welcome, since locals considered the evacuees competition for jobs
and living space. Or they were afraid of them for being radioactive.

RADIATION RISKS

Like all of the zone's residents—both permanent and occupational—
the people of Teremtsi are living models of the health impact of chronic
internal and external irradiation. They and their surroundings are re-
peatedly poked and prodded for chemical, sanitary, microbial, and radiological studies.

To check their internal cesium levels, zone residents go to the same
polyclinic that I visited in Chapter 4 and sit in the same chair to measure the gamma rays they are emitting. Strontium-90 is measured in
urine, which gives an indirect estimate of how much is left in the body.
Plutonium inhalation is estimated on the basis of the amount of air
inhaled per day, the number of hours spent outdoors annually, and the
average amounts of the isotope that get kicked up in the air. Based on
these estimates, *samosels* inhale so little plutonium that it doesn't even
influence their doses.

Despite the clean surroundings, the Ilyenoks don't grow any of
their food and buy everything from the zone administration's shops-
on-wheels that visit each of the zone's populated villages twice a week,
selling produce as well as uncontaminated bread, grains, macaroni,

canned goods, sugar, tea, candy, and other merchandise that the *samosels* can't make themselves.

The Ilyenoks are unusual. Poverty forces the vast majority of zone residents to grow and gather nearly all that they eat, which gives them most of their radiation dose. They consume cesium and strontium with vegetables, fruits, meat, and fish caught in zone waterways. But their internal doses from the cesium-137 are about four times as high as those from strontium-90.

The average *samosel's* radiation doses are double or more the maximum doses allowed for the general population. On average, half of that dose is from external radiation and half is from internal. The amounts and types of radionuclides that get into zone residents' bodies—and, thus, the size of their internal doses—depend entirely on what they eat. The Ilyenoks, who don't eat anything in the zone, probably have low internal doses. *Samosels* who don't eat any game or fruits of the forest, but who do eat local produce, will get half of their internal dose from milk and a third from meat and eggs. But it's important to point out that 40 percent of the internal radiation is of Chernobyl origin. The rest is from natural radionuclides.

A similar picture is true of the several hundred thousand people who continue to live on the highly contaminated lands outside the zone, where radioactivity is between 15 and 40 curies per kilometer—far higher than in Teremtsi—and where radionuclide levels in milk and meat exceed permissible limits from 5 to 15 times. In such regions, which should have been evacuated but haven't been, poverty-stricken fatalists have the highest internal doses, while people who feel in control of their lives and who have the resources to buy uncontaminated food can reduce their Chernobyl doses to virtually zero.

After the accident and throughout the 1990s, the Ukrainian government used active measures to reduce consumption of contaminated products. But since 2000 the budget for such countermeasures has been reduced practically to zero. Bureaucratic infighting is one problem. The Ministry of Agriculture has the radiological infrastructure in the countryside, but the Ministry of Emergencies is in charge of distributing Chernobyl-related funding and it doesn't give any to the agriculture ministry.

After leaving the Ilyenoks, Rimma and I followed Kate up a hillside of tall grasses to another occupied cottage. But in contrast to the Ilyenok's immaculate house and yard, it was a shack where weeds choked the yard and a few chickens pecked at specks on the ground. Indoors, there was almost no furniture except three beds, a deep pile of dirty rags on the floor, and an overturned basket covered with a towel to warm some chicks loudly peeping away inside.

Though I mistakenly erased the family's name from my digital recorder, in my notes I referred to the elderly woman and her two adult sons as the "drunks in Teremtsi." The men were in their mid-forties, but their faces were ravaged by drink. Smoking smelly Soviet-brand filterless cigarettes on a bench in the yard, they looked older than their mother. The scene reminded me of the Nikolai Shamenkos, the elder and younger, in Tulgovichi in Belarus.

With perverse pride, their mother told me that they grow all their own food and also gather berries and mushrooms in the forest.

I didn't know if they were apathetic, fatalistic, hungry, or just foolish. But people like this get 90 percent of their internal dose from the wild foods. Those doses are twice as high in the autumn as in the spring because the fruits of the forest are most popular in the fall. Despite constant and persistent warnings not to, more than half of the *samosels* do eat wild foods. And the reason can't be only hunger. Zone workers, who get free meals, also like to do some illicit fishing or mushroom hunting. A sign at Chernobylinterinform warns employees about levels of radioactivity in fish and mushrooms from different zone locations. But either from apathy or from fatalism, some people insist on ignoring the warnings.

To be sure, few go to highly radioactive places like the buried village of Yaniv in the 10-kilometer zone to collect mushrooms or berries whose cesium levels are in the hundreds of thousands of becquerels—except, perhaps, to study them. Mushrooms in cleaner parts of the zone are less radioactive. A sample of Paryshiv porcinis contained 900 becquerels of cesium per kilogram. Nevertheless, the maximum allowable limit for cesium in a kilogram of mushrooms is 500 becquerels. For strontium-90, the maximum ranges from 5 to 20 becquerels, depending on the type of food, because of the radionuclide's greater health risks compared to cesium-137.

Food preparation can affect its radioactivity. Drying foods concentrates radionuclides. If a kilogram of fresh blueberries contains 20,000 to 30,000 becquerels, air-dried blueberries contain more than 10 times as much (350,000 to 450,000).

Cooking, on the other hand, leaches out radionuclides. Boiling mushrooms in salt water for just five minutes reduces their cesium levels by 70 percent, while 20 minutes of boiling will reduce it by 90 percent and more. So, a kilogram of raw porcinis containing 49,000 becquerels of radioactive cesium will have only 3,400 left after boiling. It probably won't have much in the way of nutrients or taste left either, but that's a different matter.

Nevertheless, 3,400 becquerels is too high to be safe.

But how safe is safe? The answer goes to the heart of the debate over the health effects of low, but chronic, radiation doses. There is little debate that high doses have what are called "deterministic" effects—in which the same harm occurs in all individuals who are exposed to a certain dose. If you and I are exposed to a full body dose of 500 rem, we are both certain to, at the very least, develop acute radiation illness. But low levels of radiation have so-called stochastic effects, which is a fancy way of saying "random." A single alpha or beta particle or gamma ray can damage a single cell; this could lead to disease or have a hereditary effect, or it might have no effect at all. There is no way to predict which will actually happen.

It is similar to the principle that it is impossible to know when a particular radioactive atom in a bunch of radioactive atoms will decay, but it is entirely possible to know when half of the bunch will decay. Stochastic health effects can't be predicted individually, but they can be predicted statistically and epidemiologically.

The main sources of these statistical predictions are the studies of 87,000 Japanese atomic bomb survivors. Among them, an increase in leukemia was observed only a few years after A-bomb exposure. Decades later, excess cases of solid cancers were seen in lungs, breast, and thyroid, although radiation evidently does not increase the risk of cancer of the prostate, pancreas, uterus, or kidney. More recently, noncancerous illnesses such as heart disease have also been linked to radiation exposure.

From the atomic bomb survivors, scientists developed risk factors that were supposed to predict the number of "extra" cancer cases that would occur when a large population was exposed to a certain total amount of radiation, which is known as the "collective dose." The formulas estimating from 1.25 to 2.3 extra cancers for every 10,000 rem were behind the confusing and alarming predictions of thousands of "extra" Chernobyl cancers that scientists bandied about soon after the disaster. While the Soviets predicted 6,500 extra cancers among the 75 million people who lived in the European part of the Soviet Union most affected by radiation, and other experts predicted twice that number, all agreed that those "extra" cancers would be impossible to detect against the background noise of 9.5 million non-Chernobyl cancers in that same population. With the exception of thyroid cancer, as we shall see, proving that any particular cancer was related to Chernobyl is practically impossible.

But many of the initial predictions turned out to be simply wrong. Despite virtually universal expert expectations based on A-bomb research, there has not been any statistically detectable increase in leukemia—even among the 800,000 cleanup workers, or "liquidators," who were exposed to the highest radiation levels and whose health has been followed closely since the disaster.

This may be in part because many people who were granted liquidator status, and the perks that come with it, never actually worked at Chernobyl. Indeed, poverty led many people to claim Chernobyl benefits with no factual reason for doing so because official recognition as a victim meant access to income and health care. And many of those people really do believe that all of their medical problems, real or imagined, are results of the disaster. The exaggerated awareness of ill health and sense of dependence has created what some experts call the "Chernobyl accident victim syndrome."

Given more than 40 types of Chernobyl benefits, some residents of contaminated areas in Belarus can be eligible for quite large sums of money. For example, working single mothers raising children under the age of three in contaminated districts get about $30 a month—which is around one-third the average Belarusian wage and which may help explain the rise in birth rates in contaminated regions.

It may also have something to do with the rise in infant mortality in contaminated regions in the late 1990s—contrary to the trend of lower infant mortality in the rest of the country. More study is needed to figure out if this is due to radiation, poor health, and poor social services in contaminated regions or to the outmigration of educated young people. But the system of Chernobyl benefits gives single women who may lack the education, health, skills, or desire for parenting an incentive to have children and raise them in contaminated lands.

Nevertheless, most of the benefits are a pittance. In Belarus a person living in a contaminated area gets about $1.50 a month to buy clean food. The sums are so small because Belarus, Russia, and Ukraine continued the Soviet practice of giving everyone a little bit for their exposure to radiation risks rather than providing more robust compensation to people who develop actual health problems. With a total of nearly 6 million Chernobyl victims in various categories (Table 1), these puny payments add up to enormous totals.

It is a stupid system that encourages allocating scarce resources not on the basis of medical need but on the ability of a person to get registered as a victim. It also makes it difficult to perform long-term health studies since the victims' registers are clogged with fakes who skew results.

In any case, another possible explanation for the absence of leukemia is that it is incorrect to extrapolate health effects from the atomic bomb doses. Those were instantaneous and large—from hundreds to thousands of rem.

But with the exception of hundreds of firefighters who received bomb-magnitude doses, the vast majority of people affected by Chernobyl were exposed to chronic, low-level, long-term doses.

Although estimating doses in hindsight is notoriously difficult, es-

TABLE 1 Number of People Affected by Chernobyl
(to December 2000)

	Belarus	Russia	Ukraine	Total
Resettled people	135,000	52,400	163,000	350,400
People living in contaminated territories	1,571,000	1,788,600	1,140,813	4,500,413
Liquidators, 1986-1987	70,371	160,000	61,873	292,244
Liquidators	37,439	40,000	488,963	566,402
Invalids	9,343	50,000	88,931	148,274
Total	1,823,153	2,091,000	1,943,580	5,857,733

pecially low doses with no immediate health effects, one way to do so is to look for free radicals in the crystal lattice of tooth enamel. Radiation exposure creates carbon dioxide radicals out of certain impurities in the crystal. Unfortunately, a tooth must be sacrificed to do the test.

Based on these and other methods, the average evacuee from the 30-kilometer zone was exposed to an estimated 12 rem, while the vast majority of liquidators' exposure didn't exceed 25 rem. Other sources estimate the average evacuee dose at a much lower 4.5 rem. But whatever the actual figures, which are unlikely ever to be known, not only are they much lower than the doses in Hiroshima, but they were delivered over a more protracted period.

Unfortunately, there is no universal agreement on the nature and scale of health risks from such low and chronic doses. Some scientists maintain that there is a threshold dose, below which there will be no damage to health. A few even claim that low doses can have a salutary effect on health on the theory that something that's poisonous in large doses—such as table salt, magnesium, cobalt, or radiation—is absolutely necessary in small amounts.

The mainstream view is that there is no harmless amount of ionizing radiation. No matter how small the dose, all it takes is for one quantum particle to hit a molecule of DNA to cause a mutation that can lead to stochastic health effects. But at doses of less than 100 rem over a lifetime, such effects will be impossible to detect statistically. So, if you and I are each exposed to 43 microrem of radiation—the hourly average in the town of Chornobyl—there is no way of knowing if either of us will develop cancer and, if we do, whether that cancer was caused by the radiation, the fact that I was a pack-a-day smoker for more than 20 years, or the fact that you were an airplane pilot whose exposure to cosmic rays was much higher than that of a fearful flier like me.

Moreover, it seems impossible to tease the health effects of radiation out of the tangle of poverty, alcoholism, smoking, poor diet, and other factors that plague public health in places in the former Soviet Union that were unaffected by Chernobyl and that have made life expectancy—especially among men—the lowest in Europe.

It is also difficult to compare Chernobyl populations, who are poked and prodded annually, to members of the general public who don't undergo regular screening. According to some studies, for example, liquidators have higher rates of heart problems than the population at large. This could be because they inhaled the more short-lived radionuclides such as barium, cerium, and ruthenium that were more abundant during the cleanup period. Their accumulation in the lungs would have given higher radiation doses to the heart and breast—which may explain anecdotal evidence of an increase in breast cancer in young women and women who were breast-feeding at the time of their exposure.

However, critics of such studies say that the proper people with whom to compare liquidators are those who get the same amount of medical screening because such tests will turn up heart problems that would otherwise never get diagnosed. The same is true of breast cancer rates in affected areas. Breast cancer screening was barely practiced in Soviet times. Mammograms are rare to this day.

With hyperbolic interest groups trying to either exaggerate or downplay the disaster's effects, it's little wonder that Chernobyl's long-term health effects remain so controversial, allowing the nuclear industry to claim limited consequences while some politicians, activists, and victims claim a profoundly negative impact on health. The conflicting information provokes such mistrust that in 2002 the United Nations proposed creating an independent Chernobyl Research Board to design and assess research.

Journalists, both domestic and foreign, fuel the fire with their macabre tendency to focus on sensationally deformed children even if they were born far from Chernobyl and their maladies cannot be traced to the disaster. In fact, the descendants of A-bomb survivors have shown no increase in congenital deformities and the same is true of Chernobyl survivors. What deformities occur are those that, sadly, occur in any population.

The sins of the other side are more difficult to demonstrate because they are less graphic and dramatic. But nuclear proponents' assertions that the only proven health impact has been an increase of childhood thyroid cancers are also misleading.

Unfortunately, there have simply been too few properly designed, impartial medical studies to prove much of anything about the disaster's long-term health impact. One thing is certain, however. The nuclear industry—both in Ukraine and internationally—has done little to advance Chernobyl health research. In fact, even after the disaster, the Chernobyl nuclear power plant was quite a profitable enterprise until it was shut down in 2000. Throughout the 1990s, the company town Slavutich spent lavishly on festivals and free concerts with expensive foreign performers.

In all that time, however, the plant did not provide a penny for cleanup works, scientific research, health care, resettlement costs, or compensation in connection with the disaster. Nor did it pay anything for the administration of the exclusion zone.

CHERNOBYL'S CHILDREN

In sharp contrast to the matriarchy of *samosels,* men dominate the zone's working population. Women have made up no more than a third, and usually significantly less, since the disaster.

The zone's labor force is a demographically unique group. Since it is not a permanent population, there are no birth statistics and no natural population growth—with one exception. In 1999 a baby girl was born to zone workers who set up housekeeping in an abandoned house in Chornobyl because they had no other place to live. Since pregnant workers are not allowed to work in the zone, the woman hid her condition until the very last moment and gave birth at home with her partner's help. As of 2004, however, five-year-old Maria was fast approaching school age and the unique little nuclear (as it were) family was waiting for the government to provide them with housing outside the zone.

Little Chernobyl Maria, as she became nicknamed, was blithely unaware of the controversies and arguments her very existence sparked among zone workers. Though outsiders were touched by Maria's story, most people in the zone thought that the little girl's parents were exploiting her to extort bigger and better housing, especially after they turned down an apartment in a small town nearby and insisted on a place in Kiev. Certainly, they were exposing her vulnerable growing body to more radioactivity than could possibly be good for her, even if it was to cesium and strontium rather than the radioactive iodine that caused an epidemic of thyroid cancers in Chernobyl's aftermath.

In direct contrast to the experts' overestimating the dangers of post-Chernobyl leukemia, they certainly underestimated the threat of thyroid cancers. Indeed, when reports of a rise in thyroid cancers first surfaced in 1991, the international community initially treated them skeptically. It was thought that extensive screening programs in Ukraine and Belarus were simply leading to earlier detection. The cancers, said the experts, were surfacing too early since the latency period for childhood thyroid cancer was believed to be 10 years; prior to Chernobyl, there had been no reports of thyroid cancers earlier than 5 years after exposure.

But those predictions were based on very limited studies. Prior to Chernobyl, the largest study of childhood thyroid cancer involved only 58 patients.

Childhood thyroid cancer is an extremely rare disease, so any increase in the number of cases is noticeable. In the decade before the disaster, there were a total of seven cases in Belarus. But between 1986 and 1998, there were more than 600. Most of them were concentrated in the Gomel and Brest regions, although no one has yet figured out why there are more cases in Brest than in Mogilev, which was much more contaminated. It could be because iodine deficiency was more prevalent in Brest, which lies near the Polish border. In the USSR, stable iodine was added to bread rather than salt, but a resurgence of endemic goiter in Belarus indicated that iodine supplementation had come to an end in the early 1980s.

Whatever the reason for the geographic anomaly, accusations that the thyroid cancers could have been averted if the Soviets had systematically distributed stable potassium iodide (KI) tablets after the disaster may not be entirely true. Taken *before* exposure, or within hours afterwards, KI protects against packing the thyroid with radioactive iodine. But taken too much time *after* exposure, KI may actually "lock in" the radioactive isotope and accentuate the dose. In fact, KI may itself be carcinogenic.

The best solution probably would have been to distribute powdered milk, so that children would not have consumed the radioactive milk from cows grazing on grasses coated with radioactive iodine. One study in the Bryansk region of Russia found a direct link between a child's consumption of milk and dairy products in the first few months after the accident and the risk of thyroid cancer. The more milk, the higher the dose of radioactive iodine and the greater the risk to the child.

As of 2002, about 1,800 cases of childhood thyroid cancer had been traced definitely to Chernobyl. About 60 percent of cases were in Belarus, 40 percent in Ukraine, and 3 percent in Russia. Conservative estimates of the total number of thyroid cancers that will occur over the lifetimes of people who were children in 1986 is between 6,000 and

8,000 in the three countries. Although thyroid cancer can be treated, all of these people will need to take thyroid hormones their entire lives and be under constant medical supervision.

Other than dose, a child's age is a very important factor. The younger the child in 1986, the higher was the risk of developing thyroid cancer. By 2001 the cancer risk in children who were older than 10 appeared to be tapering off, while it continued to grow for those who were younger than 5. In children born after the disaster, thyroid cancer incidence is similar to predisaster rates. It is likely that the coming decades will also see an increase in other solid cancers; however there is no consensus about the number of cases or types of cancers that will occur.

Cancers are not the only health problems, although tracing the causes of noncancerous illnesses to radiation is more controversial. The congenital heart disease that was the heartbreaking subject of the Oscar-winning documentary *Chernobyl Heart* is one example of a disease whose occurrence cannot be linked definitely to the disaster, although it might be.

Also, high rates of respiratory illnesses in contaminated regions may be due to long-term exposure to low doses of cesium-137, which lowers the number of T-cells and other immune cells in children. Such immune deficiencies have come to be called "Chernobyl AIDS" since AIDS also destroys T-cells. Some similar studies have been criticized because they compared public health in Chernobyl-affected regions today with the same regions prior to 1986, which failed to account for the myriad other social, economic, and political factors influencing public health in the former Soviet Union.

The plight of Chernobyl's children has created an international humanitarian industry, whose value has yet to receive detailed study. Perhaps the highest profile are the voluntary initiatives supporting "health holidays" for Chernobyl children in other countries such as Cuba, Ireland, Canada, and the United States. The United Nations called them "possibly the largest and most sustained international voluntary welfare program in human history." A total of several hundred thousand children have taken part since the accident, though there is no consensus in the international health community as to the claimed benefits of "detoxification."

While most programs' promotional materials are rather fuzzy on the details, the general idea seems to be that by spending a summer in a noncontaminated environment, the children have an opportunity to cleanse their systems of radionuclides. But while this may be true of cesium, which can be cleansed from the body in three months, it doesn't work for strontium built up in bones or for the radioactive iodine that is long gone but whose health impact is the greatest. In fact, since most health holidays don't last more than two months, there isn't even enough time to clean out cesium.

Unfortunately, many of these well-intentioned programs are fueled by bad or simply misleading science. For example, the Chabad Children of Chernobyl project claims to be saving the lives of children from thyroid cancer by "evacuating" them for permanent residence to Israel. Leaving aside the hyperbole of using the word "evacuate" many years after the disaster, Chabad goes so far as to maintain that children born after the disaster have higher rates of thyroid problems than those who were children at the time, which is just blatantly false. Nevertheless, Chabad is hardly alone in trying to garner support for its activities with misleading pseudoscience. Nearly all of the Children of Chernobyl public relations materials do the same.

One problem is the dilute definition of "Chernobyl children." Chabad—which finds the kids by running radio ads offering to take Jewish children to Israel for free education and medical care—says that "any child living within the Chernobyl area is eligible. Sick children are given priority, but every child living within the contaminated area is considered to be at high risk."

If the "Chernobyl area" includes lands contaminated by at least one curie of cesium for each square kilometer, their population totals 4.5 million people, including about 1 million children. In a world of unlimited resources, it would perhaps be right to treat all of these children as equal victims. But in the real world it seems profoundly unfair. A child living in an area contaminated with 1 curie is simply not at the same risk as one surrounded by 39 curies. But Chabad and the health vacation charities make no distinction between them. They also make no distinction between children who are exposed to radiation and those who have actually suffered health effects.

Since this is pretty much the same philosophy that the Soviets used in distributing Chernobyl compensation, the result is the same as well—everyone gets a few pennies, which means that there is a lack of funding to help the truly ill.

The people who take children on health vacations or help them immigrate permanently to Israel are clearly well intentioned and charitable. But it seems a monumental waste of money that could so much be better spent on basic medical infrastructure in the affected areas (and in monitoring the proper use of the money, to which too little attention is paid). It also has untoward psychological effects, since their well-intentioned hosts often treat the children as victims and expose them to a standard of living they have little hopes of acquiring at home.

Most Chernobyl charities also engage in less expensive but more effective forms of aid such as delivering medical supplies, sponsoring families, and funding local nurses, although the health holidays remain popular.

Here are a few facts and figures:

* What the Irish Chernobyl Children's Project spent on health vacations in 1991: $10 million
* What Chabad has spent on resettling 2,000 children to Israel: $30 million
* What Ukraine spent on health care for Chernobyl victims in 2000: $6 million
* What the average Belarusian earned in 2000: $700
* What the average Ukrainian earned in 2000: $1,000
* What it costs to provide one Belarusian child with a health vacation in California: $1,200.

A young doctor when she was sent to Chernobyl in 1986, Olia Senyuk thinks that she lost her thyroid gland to cancer as a result, although the prevailing medical wisdom is that radioiodine poses dramatically lower risks to adults' thyroids than to children's. Children are considered more vulnerable because their thyroids are smaller than adults', and therefore a greater proportion of the gland is exposed to radiation. But studies of liquidators—including doctors like Senyuk—show increased incidence of thyroid problems, although this could be an artifact of screening turning up diseases that might have otherwise been missed.

In general, the health of liquidators is worse than that of the general population, though it is difficult to separate the effects of radiation from depression, alcohol and tobacco consumption, and other factors. Liquidators, for example, have among the highest rates of suicide in Ukraine.

Their children may also be at higher risk, but the evidence is ambiguous. In one study, children conceived within two months of their fathers' work on the cleanup had a higher frequency of certain genetic mutations than those conceived after four months and more. But this has not yet had an evident impact on their health. Experiments with mice suggest that leukemia risk can skip a generation, but there has been no evidence of this in the children of Chernobyl liquidators.

Rimma Kyselytsia introduced me to Senyuk in a laboratory tucked into a corner of one of the zone's scientific centers. The necklace scar plainly visible above the collar of her blue blouse didn't seem to trouble her, nor did the fact that she was still working in the zone 18 years later, studying the DNA in the white blood cells known as leukocytes donated by people who work at the Sarcophagus covering the ruined fourth reactor. She told us that chronic exposure to low doses of radiation has effects similar to those of aging and that Shelter workers get cardiovascular diseases at a younger age than the rest of the population.

"People who have accumulated large radiation doses have more sensitive DNA that breaks at lower radiation doses than those required to break the DNA of people who have accumulated smaller doses," she said. "At the same time, the people with large doses have DNA repair mechanisms that are, in a sense, well trained. So, even if their DNA breaks more easily, it is also repaired more easily."

Senyuk's confident contention is not exactly conventional wisdom in radiology circles, but it is not far-fetched either.

Nevertheless, a cell's DNA mechanics can make mistakes, especially if both strands of the DNA molecule are broken rather than just one strand. If these mistakes occur in the so-called junk DNA that makes up the vast majority of the human genome, nothing may happen because junk DNA doesn't hold the codes for anything useful. If the mistake is in a gene, which codes for proteins—the building blocks of life—the result can be a defective protein. If the mistake happens in the tumor suppressor gene TP53, the result can be cancer because TP53 codes for a protein involved in apoptosis, which makes damaged cells die before they cause trouble. If the protein is defective and the damaged cells live, they can go on to become malignant. This, at least, is what has been observed in rodents. There is little direct evidence about how radiation causes tumors in people.

In this light, it was all the more surprising to hear Senyuk say that Sarcophagus workers, who have higher radiation exposures than just about any other professional category in the zone, are generally in better health than people who work in less radioactive areas. "People now working at the Shelter have undergone selection," she explained. "Weaker people either died or their health worsened so that they can no longer work in the zone. The ones who remained are more radiation resistant."

When it comes to radiation, not all of us are created equal. About half of the population is average, a quarter is very sensitive, and the other quarter is very resistant. A dose that will not have any effect on the average person can have a negative impact on the very sensitive, while a dose that damages an average person, will not harm a radiation-resistant one. Senyuk was working on developing supplements called radioprotectors to strengthen radiation resistance in the weak by increasing the dose needed to elicit an effect.

Before we left, Rimma asked if hay fever could be caused by radiation exposure. For the first time in her life, Rimma developed hay fever and high blood pressure that spring.

"Sure," said Senyuk. "Allergies, diabetes, general malaise, and fatigue."

"Then I want to volunteer for your experiments, too," Rimma said firmly, promising to call Senyuk soon to pick up whatever concoction she was testing at the time.

Nevertheless, the connection between low doses of radiation and medical conditions, such as allergies, that can have a multitude of causes is difficult to prove.

BRUNCH WITH THE DEAD

The zone is a strange place on an average day, but one of its most bizarre sights can be seen only on the Sunday after Easter. And it is bizarre precisely because—at first glance, at least—it seems so normal. The town of Chornobyl is crowded with people, you can hear voices other than your own in Pripyat, and there are actually other cars on the roads. About half of the zone's 30,000 annual visitors come at this time.

It is also the only time of year when you can see teenagers in the zone, though children are snuck in occasionally to visit with their *samosel* grandparents.

Exceptions to the rules are made on the Sunday after Easter because this is when everyone traditionally goes to the graveyards for lunch (and many, many drinks) with the dead. A vestige of pagan ancestor worship, masked with a thin veneer of Christian ritual, the holiday is known as *provody* or "bidding farewell." I had attended my share of Ukrainian-American *provody* in the United States, but always found it exceedingly weird to see people laughing, picnicking, and drinking on graves—at least when I was sober. Impaired sobriety does much to get into the spirit—so to speak—of things.

I had no hopes of impairing my sobriety on *provody* in the zone since I was behind the wheel. It was a gorgeous, sunny day in mid-April. The air carried notes of early spring as I drove with open windows along the cracked and potholed road that led from Chornobyl to the western border of the 30-kilometer zone. Usually, it is advisable to keep windows closed. Although a moving car stays ahead of the dust it kicks up, it gets enveloped in the wake as soon as it stops. But it had rained the previous day and there was no dust.

My Chernobylinterinform escort for that day was Serhiy Chernov, a gruff, mustached, 40-something dressed in the usual camouflage. We were on our way to Lubianka, a village on the floodplain of the Ilya River. Before the evacuation, it was home to 700 people. In 2004 there were 17. But on *provody* there were likely to be many more people visiting the cemetery. Of the Ukrainian zone's 15 inhabited towns and villages, Lubianka was one of the most contaminated—though in Belarus, Tulgovichi was much "hotter." Colored orange on the radiation maps, each square kilometer of Lubianka was contaminated with 20 to 50 curies of cesium-137 in 1986. The Ilya, a small tributary of the Uzh River that originates in the swamps of Belarus, was the most strontium-contaminated waterway in the zone in the immediate aftermath of the disaster, although the river's current had largely cleansed it.

After driving for about an hour, we crossed a bridge over the Ilya and turned onto the gutted and gouged dirt road that led into Lubianka, passing a pasture of cows and horses. Lubianka had more livestock than any inhabited village in the zone. Of the zone's population in 2004 of 41 cattle, 17 lived in Lubianka. Nearly 30 pigs, 6 horses, and 11 goats were also scattered among the zone's *samosel* households.

Our plan was to find Lubianka's "mayor," Oleksandr Tkachenko. Serhiy had met him a year earlier when a group of latter-day Cossacks visited Lubianka to paint its World War II statue, chop wood for the winter, and proclaim Tkachenko the mayor. Tkachenko was also unusual in that he was one of the very few zone residents who was male, working age, and sober. Two-thirds of the zone's population is women, nearly all of them pensioners. Most of the men are elderly as well, although the handful of younger ones are troubled in one way or another, like the drunks in Teremtsi.

After parking in a clearing where only birdsong broke the silence, we passed some decrepit and overgrown cottages before coming upon the Tkachenko's green plank fence. Two paper handbills warned residents that spring was a dangerous season for fires. The gate was open and, ignoring the barking gray mongrel chained in the yard, Serhiy called Tkachenko's name.

His wife, Maria Shevchenko, emerged from the house, wiping her hands on a smudged apron, and Tkachenko soon followed. She was short and plump. He was tall and lean. And they were just preparing to walk over to the graveyard but graciously agreed to show me their homestead (Plate 6).

After leading us through a yard of cats, chickens, and rusty metal sheds, they showed us the small plowed field where they grow nearly everything they eat. My dosimeter displayed levels around 25 to 30 micros. This was on the high side. Most zone residents live on land where background readings don't exceed 20 microroentgens an hour.

A square meter of Lubianka soil contains as much as 400,000 becquerels of cesium, 125,000 becquerels of strontium, 4,000 becquerels of plutonium isotopes, and almost as much americium. Only two other inhabited places are dirtier: Zalissia, just outside the town of Chornobyl, where the cesium content per square meter of soil is as high as 330,000 becquerels and the strontium content is even higher than Lubianka's. Radiation exposure runs up to 50 microroentgens an hour in places.

But Novoshepelychi, the sole inhabited place inside the 10-kilometer zone, exceeds all permissible limits. Ukrainian law mandates resettlement from places contaminated with 555,000 becquerels of cesium, 111,000 becquerels of strontium, and 3,700 becquerels of plutonium in a square meter of soil. That amount of Novoshepelychi soil contains up to 1.7 million becquerels of cesium, 750,000 of strontium, 13,000 of plutonium, and nearly as much americium!

From 1987 to 2000, Novoshepelychi was the site of an experimental farm where scientists bred cattle from a bull nicknamed Uranium that had escaped the evacuation of livestock after the disaster and was captured together with three dairy cows: Alpha, Beta, and Gamma. Uranium was the grand sire of four generations numbering 200 descendants before he died in 2001 at the age of 15.

Sava Obrazhei and his wife Olena had returned to their old home in Novoshepelychi and asked for a job tending the cattle, which the scientists agreed to because none of them had any intention of living in the highly radioactive village full-time. But when the farm moved to less contaminated land, the Obrazheis were left without jobs or electricity. The zone administration had decided that it was too expensive to maintain electric lines to villages with only one or two people, so it encouraged lone residents to move to other zone villages with more people. But the Obrazheis refused to leave and had been living without electricity since 2001. They were by no means poor, however, at least by zone standards. Compared to the average monthly pension of about $20, the Obrazheis were wealthy, together receiving $150!

I had visited them in 2002, just a few weeks after one of their adult sons was attacked and drowned in the Sakhan River that runs alongside their lands, probably a victim of poachers or outlaws. Another son had been killed a year earlier under still more mysterious circumstances. Most people thought it very strange that they stayed on after such tragedies. But Sava and Olena continued to plant vegetables in soil where Geiger counters measured more than 130 microroentgens an hour—even hotter than Tulgovichi in Belarus.

After the brief tour of their land, the mayor and his wife invited us inside their modest cottage. Typical of zone dwellings, it had one story and two rooms that smelled a bit musty. The small windows let in little natural light, which was also blocked by the budding fruit trees in the yard. But it was very clean and decorated with colorful blankets and pillows that Maria embroidered in the winter, when there was no fieldwork to be done. I didn't ask why there was a small inlaid wood portrait of Lenin on the television and a religious icon on the wall above it. Their ideologically eclectic décor was their own business.

Like most zone residents, the couple cooked and heated their home with wood-burning ovens that didn't do a very good job of keeping them warm in winter because the cottages are not well insulated or sufficiently winterized. A zone home's average winter temperature is 15°C (59°F).

Of course, Wormwood Forest wood contains cesium and strontium. But the radionuclides are spread through the large volume of the wood. When it is burned, however, the part left behind as ash is about 10 percent of the original volume—yet it contains the same amount of radionuclides. This concentration of radioactivity is what makes wood ash one of the most dangerous sources of exposure for zone residents. The maximum permissible limit of beta radiation is 20 beta particles emitted from a square centimeter in a minute. A square centimeter of ash in zone stoves emits from 28 to 315 beta particles per minute. In Lubianka the rate is as high as 47 beta particles per minute.

Since zone residents use their stoves all the time—for cooking year-round and for heating in colder seasons—they are vulnerable to breathing in particles of ash, which can lodge strontium in their lungs. If it gets on their clothes or in their shoes, it can expose the skin to radiation.

Getting dirty with contaminated soil and ash is why *samosels* themselves can be sources of radiation. A square centimeter of the average zone resident's skin and clothes emits 22 beta particles per minute, but this is an average. Some zone residents exceed it; others don't. The rates have also decreased significantly in the nearly two decades since the disaster. In 1989, residents of a village at the Pripyat delta that was a shade less contaminated than Lubianka wore clothes that emitted 60 beta particles a minute.

Carrying plastic bags of food and drink, Tkachenko and Shevchenko piled into the back seat of my car and directed me onto a path of packed sand that we bumped along for a few minutes before coming upon dozens of cars and buses at the foot of a hill.

After parking the car near a grimy old bus, I followed my passengers up the path to the cemetery and pulled out my dosimeter. It quickly beeped out a reading of 100 microroentgens on the path before dropping down to around 50 when we turned onto grass. While my companions unpacked their brunch on a picnic table near a Tkachenko family gravesite, I made my leave to wander around with my notebook and dosimeter.

There was nothing about *provody* in the zone that was not strange. More vehicles were parked by the Lubianka graveyard than I had seen in all of my zone visits combined. Although there were some old wooden crosses that had rotted at their base and fallen over on the ground, most of the grave markers were Soviet-era crosses made of welded pipe. All of them were dressed for their annual party with wreaths of plastic flowers, ribbons, and colorful scarves that gave the cemetery a festive air.

There were easily 150 people in the cemetery, and the droning sounds of distant conversation and laughter gave it a sense of life that was missing in the village itself. Some people sat right on the graves, but larger families usually gathered around picnic tables, making sure to include the dead by spreading a tablecloth over the grave and piling it with the deceased's favorite foods as well as the occasional bottle of vodka and shot glass. But even on graves whose overgrowth of weeds and crumbling markers indicated an absence of recent visitors, someone had placed a single candy.

Seeing me with a dosimeter and an evidently lost look, Nadia Artemenko invited me to join her and her sisters who had been evacuated to a village just outside the 30-kilometer zone's border. But when that village proved none too clean, they were moved to another village, also named Lubianka, south of Kiev.

The sisters had come to the original Lubianka on a bus provided by the zone administration to visit the grave of their mother, who died in Lubianka in 1989. Her grave was blanketed with ground-hugging evergreen rosettes that symbolized eternal life. The plants were piled with pink and white plastic lilies, and a collection of gardening tools used to tend to the grave lay in a pile nearby.

"My mother came back here from the evacuation two years before she died," Nadia explained to me, tucking a wisp of white hair back under her red paisley kerchief. "She wanted to die in her home."

I barely sat down before the women started plying me with food: a fried drumstick, roast pork, sausage, homemade apple cider. Although I wasn't hungry, it is rude to refuse food, so I accepted some cider—which was quite good—and a chunk of sausage.

The same thing happened with every group I visited: offers of food mixed with so many tales of death and loss, evacuation and fear, that they all began to run together—the man whose wife died in a plane crash soon after they were married; the young nephew who died of cancer; the fathers, brothers, uncles, and cousins who died in World War II. But no one to whom I spoke had lost anyone as a direct result of the Chernobyl disaster.

Trying to avoid any more food that it would be rude to refuse, I looked for someone who was doing something other than eating and came upon Motrona Shevchenko weeding her parents' graves inside a wrought iron enclosure. The graveyard was full of Shevchenkos and Yushchenkos, with a smattering of other surnames. Evidently, Motrona was a popular name for women in Polissia. I've never met a Motrona in Kiev, but I had already met two in the zone.

Under a black kerchief, Motrona's face seemed frozen in sadness and she started to cry as soon as I asked her if we could talk.

"The whole family gets together for *provody*," she said, gesturing at the 15 people gathered around an adjacent picnic table. "It is a time of joy that we're together, and sadness for those no longer with us."

Her face streaked with tears, she continued puttering around the graves and I tried to leave her alone with her grief, but she couldn't let me go without feeding me. Despite my protests, she stuffed a cellophane-wrapped chocolate cake in the deep pocket of my camouflage jacket.

Not wishing to intrude on any more private rituals, I found an empty picnic table and sat there to make notes. It was almost three o'clock and the distant buses were honking that it was soon time to leave. Paraska Shevchenko, a spry 74-year-old, carried a tattered plastic bag full of leftovers from her *provody* lunch and put it on the table behind me while she stopped for a rest. Like everyone in Lubianka, Paraska had been evacuated to Lubianka-2.

"But I sneaked back in to work on my garden," she confessed. "We had already done our spring planting and I wasn't going to just leave everything untended."

Paraska noticed my dosimeter beeping by my side. "What's it say?"

I showed her the digital display reading 59, 66, 50.

"Is that milli," she asked.

"Micro," I responded.

"That's nothing," she said with a dismissive wave.

Paraska motioned at the graveyard. "Had you come here for *provody* in the early years after the evacuation, this place would have been much more crowded. But many people have died."

She sighed. "And there are those who want to forget that they were ever here."

The distant honking continued and Paraska left to catch the bus. People were drifting slowly out of the graveyard. Car doors slammed and engines revved amid the sounds of parting and farewell. Soon enough, the only sounds in the graveyard were the wind in the pines and the spring song of chaffinches.

Thinking that my escort, Serhiy, might be worried about my whereabouts, I wandered over to Tkachenko's table, where I added Motrona's chocolate cake to the mortuary feast.

The mayor and his guests were among the last people left, except for two militiamen who had been standing outside the cemetery since we arrived. The zone administration always makes sure that the fire department is on alert and that there are law enforcement officers at all of the graveyards. Past cemetery revelers started several *provody*-related fires.

Tkachenko poured Serhiy a shot of his home brew for the road (or, as they say in Ukraine, "for the horse"). Because *samosels* make their moonshine from sugar, which they purchase from the mobile shops, it is usually not a radiation concern—though the taste reminds me of my college chemistry lab. But when Tkachenko explained that his secret ingredient was birch sap, I was glad of my designated-driver status. A popular beverage in Ukraine, birch sap also concentrates radionuclides. A liter of the stuff in one 2003 sampling contained 1,800 becquerels of cesium. Birch sap, together with things such as nuts, falls under the "other" category in Ukraine's list of radioactivity limits in different foods. The maximum permissible cesium level in a kilogram or liter of "other" is 600 becquerels.

Meanwhile, Shevchenko busied herself setting out eggs, cucumbers, and slices of bread. After refusing so much food during my *provody* mingling, I was actually hungry but wary about the source of the victuals. While the cesium in Lubianka's homegrown produce usually doesn't exceed permitted radioactivity levels, milk and eggs do. Indeed, in 2002, cesium levels in Lubianka milk were almost seven times as high as the maximum allowable limit of 100 becquerels per liter. They were the highest of any inhabited place in the Ukrainian zone, although Tulgovichi in Belarus probably held the record. Moreover, if the permissible level of strontium-90 in a kilogram of vegetables is 20 becquerels, Lubianka's measured as much as 90. But Lubianka is not the leader when it comes to strontium in vegetables. This honor goes to the village of Zalissia, just outside the town of Chornobyl, where some vegetables measured 420 becquerels of strontium.

Seeing my hesitation, Shevchenko told me that all of the food was store bought: "It's too early in the season for our own produce." Thus assured, I munched on a cucumber spear and listened to Tkachenko complain to Serhiy about his zone permit, which identified him legally as a resident of the town of Ivankiv, about 10 miles south of the zone.

"That's insulting. This is my home. Why don't they just write 'resident of Lubianka'?" he asked adamantly and then, flashing his gold-faceted smile, added: "Or better yet, 'mayor.'"

None of us had a ready answer. The *samosels* pose one of the zone's most vexing problems, but they are a problem that will most likely go away with time. Sadly but inevitably, they are simply dying off.

While my hosts packed up the remains of the picnic, I decided to take one last stroll around the graveyard. Motrona was tidying the graves of some other Shevchenkos, but there were almost no other people left. As I wandered around the festooned and forlorn gravesites, listening to frogs trilling in a nearby swamp and the croak of a great reed warbler, it occurred to me that when the last Motronas die and there is no one to remember the zone's dead, nature will still sing their requiem. The thought was both sad and comforting.

8

The Nature of the Beast

*Windows are boarded, and a padlock hangs like a rusted earring
on the fang of a doorlatch.*

Lina Kostenko

T
he elephant whose foot is at the very center of the Zone of
Alienation is a unique and exotic beast, but it isn't an animal.
"Elephant's foot" is the nickname given to one of the globular
masses formed by the melted reactor core once it cooled in the
disaster's aftermath. The elephant's foot, together with nearly 200 tons
of nuclear fuel and fission products that exploded, burned, melted,
and poured into the nooks and crannies of the demolished reactor
building, have been the proverbial elephant in the room of this story,
the subject I have deliberately avoided mentioning, though it is both
metaphorically and literally at the very heart of the Chernobyl disaster
and its aftermath.

Encased in the cracked and unstable Sarcophagus and so lethally
radioactive that no one can get close enough, long enough to effec-
tively study them, the elephant's foot and other fuel-containing masses
are among the zone's greatest scientific mysteries and its greatest long-
term dangers. Because in the absence of huge scientific breakthroughs
and even larger amounts of money, that nuclear debris will mar
the Earth so deeply into time that, in human terms, it may as well be
forever.

It is a difficult notion to accept. One of human culture's most
primitive fears is of contamination, conjuring violations of purity and

217

sanctity, the defilement of nature. Pollutants are out of place, violating the established order and provoking fear, disgust, and avoidance. Radiation is especially wont to provoke these primal anxieties since it is invisible to our senses, yet the very knowledge of its presence fills us with helplessness and uncertainty.

Splitting atoms, the fundamental building blocks of matter, certainly does seem to violate the established order—especially when it produces a cauldron of highly radioactive isotopes and dangerous, man-made elements such as plutonium and americium whose pollution of the Earth is practically eternal. The link between radiation and pollution has fueled arguments that nuclear energy is "unnatural," and in a very real way, it has become a totem for the artificial and man-made, an energy source that must be "contained" in sterile conditions and kept separate from the organic world.

Few objects on the planet look as unnatural as the Sarcophagus and its radioactive contents. It is a huge and ugly pollutant in the midst of the wilderness. But is that contaminated wilderness "natural"? As the British sociologist John Wills writes: "'Nature' is a difficult term to appreciate fully, its multifarious dimensions giving rise to its status as perhaps the most complex word in the English language." He coined the phrase "unnaturally natural" to describe the paradox of man-made contamination—such as that surrounding some American nuclear weapons facilities—leading to the exclusion of people from swaths of territory that then become undisturbed wild lands. It is corollary to Bruce Sterling's involuntary parks.

For if nature refers to the essential, innate quality of things, the elephant's foot and other fuel masses, the Sarcophagus, and the radionuclides that have crept into all the links of the zone's food chains have become unnaturally natural elements of the zone. The artificial has become an integral part of the natural in the radioactive wormwood forests of Chernobyl.

TECHNOGENESIS

On a sunny day in May, not long after the disaster's eighteenth anniversary in a quickly blooming spring, I drove into the 10-kilometer zone with Rimma Kyselytsia. We didn't encounter a single other car on our way to the nuclear station. The pine trees planted after the disaster had grown so tall that it was difficult at times to see the Sarcophagus on the horizon. Only its ventilation stack was visible as we passed a forest of steel towers draped with transmission lines that carried 308 billion kilowatt-hours of electricity before the last of Chernobyl's original four reactors was shut down in December 2000. But ending power generation was not the same as closing the plant. Although there was little evidence of them outdoors, thousands of people still worked there: decommissioning the reactors, building nuclear waste facilities, maintaining and monitoring the cracked and decrepit Sarcophagus.

The plant's parking lot was wet when we drove in and got out of the car.

"They washed it down," I noted. Although the contaminated asphalt was bulldozed and buried after the disaster, as were most of the roadsides, it was—and remains—impossible to clean the area completely. Wind blows contaminated dust and sand onto the asphalt, which must be sprayed regularly.

"A French journalist once thought she caught me in a lie because she had spent several days here but never saw the roads being washed," Rimma recalled. "But then I pointed out to her that it was winter and if the roads are sprayed, they will ice over."

Rimma shook her head at the memory as we walked past a silvery, larger-than-life bust of Lenin in the plaza that fronted the plant's lobby. The Soviet plant's full name was the "V. I. Lenin Chernobyl Atomic Energy Station," named after the man who once declared famously (if inscrutably): "Communism is Soviet power and the electrification of the entire country." Ukraine dropped the "V. I. Lenin" from the official name, though the plant's original signs still displayed it, like a mausoleum to Soviet times. In fact, nearly all of the old Soviet symbolism remains where it was at the moment of the explosion because it is radioactive.

Lenin's statue would surely have graced the grounds even if the plant had been named after Leonid Brezhnev, who was the USSR's increasingly incoherent general secretary when Chernobyl's No. 1 reactor went on-line in 1977. By the time the No. 4 reactor was completed in December 1983, Soviet spymaster Yuri Andropov was in charge, and when it exploded in 1986, Mikhail Gorbachev—the Soviet Union's last leader—had been in office for a year.

I didn't bother to ask why the Lenin statue was still there, long after his ideology had been knocked off its pedestal. Although a wave of Lenin deconstruction followed in the wake of the Soviet Union's

1991 collapse—including a very large monument in the center of Kiev—there were still plenty of Lenins around. A modestly sized and esthetically unremarkable Lenin still stood in downtown Kiev. Only some elderly Communists care if it is there or not.

The town of Pripyat never had a Lenin monument, although its poplar-lined main boulevard carried his name.

On the colored radiation maps, the entire four-square-kilometer grounds of the power plant (not including the cooling pond)—together with Pripyat and the Red Forest—are marked with cross-hatching that the legend describes as "territory that experienced intensive technogenic influences as a result of decontamination works." The word "technogenic" raises red underlining on my word processor's spell check and doesn't appear in my 10-pound Random House dictionary. For that matter, I couldn't find it in any Ukrainian or Russian dictionaries. It seems to be a neologism created by combining "technology" and "genesis" and, thus, can be defined as something of technological origin—unnatural, artificial, man-made.

My dosimeter beeped around 30 to 40 microroentgens an hour as we walked towards the nuclear plant's administrative building, south of the reactor complex. As soon as we entered the lobby, it dropped to 20. Farther inside the building, background was perfectly normal.

Stanislav Shekstelo, the deputy head of Chernobyl's information department, was there to meet us but had only ordered a visitor's badge for one. So, Rimma went to drink coffee in the cafeteria while I followed him through the security checkpoint and then up two flights of stairs to a large room with a model of the site and informative posters about its past, present, and future. There I learned that the nuclear plant's work force had declined from a peak of around 12,000 in 1996 to 9,050 when it closed in 2000. Projected to number 3,300 in 2008, it will continue to decrease as time goes on.

Shekstelo explained that, officially, the Chernobyl Atomic Energy Station was no longer an atomic energy station since it no longer generates energy. It was the Specialized State Enterprise "Chernobyl Atomic Energy Station," whose main task was decommissioning the reactors and transforming the Sarcophagus into an environmentally safe system.

"Environmentally safe system" was actually the official lingo, to be found in all of the plant's informational materials. Volumes, however, can and have been written about whether this is even possible and, if so, how to accomplish it.

As the Special State Enterprise's assistant director, Oleg Goloskokov was in charge of the process and he soon joined us in the model room. A tall, blue-eyed Russian who has worked at Chernobyl since 1989, Goloskokov's life has been tied to the atom since 1957. He was five years old and staying with his grandparents when the infamous Kyshtym nuclear spill forced their evacuation. Thirty villages were wiped off the map.

When I asked him his opinion about nuclear energy, he didn't say whether it was good or bad, but slashed the air with his hands, making imaginary lines and boxes as if to regiment and enclose the rules and regulations that he said must be followed with any inherently dangerous technology. But he was cheerful enough during our chat, filled as it was with depressing numbers and insoluble problems.

Perhaps the greatest problem was figuring out what exactly was going on inside the Sarcophagus or, as it is officially called, "the Shelter Object."

"We don't know how much of the fuel is still inside the reactor building," Goloskokov said. "There were about 200 tons at the time of the accident, but research has produced inconsistent results. According to various estimates, there are 160 to 180 tons remaining inside."

Since the conventional wisdom holds that about 3.5 percent—or seven tons—of the reactor fuel was expelled in the explosion and fire, even the upper estimate of 180 tons signifies quite a lot of "missing" fuel.

Part of the missing fuel might be pieces of the core that the explosion threw into the reactor's central hall. But a lot of the lead, boron, and sand that was dropped by helicopters to put out the graphite fire in 1986 actually missed the fire and fell instead on the core fragments, blanketing them with extinguishing materials that were 30 feet deep in places.

"We don't know how much fuel is underneath it and we don't know what form it's in. All we know is that it's there!" Goloskokov laughed in punctuation. At first I thought his sense of humor odd. But the more he talked, the more I realized that black humor was not a bad way to deal with the mysteries, puzzles, and ironies of Chernobyl.

The estimated 160- to 180-ton figure was not limited to the fuel that was never ejected outside the reactor building in the explosion and fire. It also included ejected radioactive debris that was excavated from the nuclear power plant's grounds together with vegetation, asphalt, and deep layers of soil. Packed into containers, the stuff was stacked into the Shelter Object's northern wall, known as the Cascade Wall because of the "steps" that cascade down the structure's side. In the photograph of the Sarcophagus in Plate 7, the Cascade Wall is on the left.

No one knew how much fuel was mixed up with the Cascade Wall's contents, but at least they knew where it was. Far less was known about how much of the fuel was underneath the sand, gravel, concrete, and other building materials used to reconstruct the plant's grounds after the decontamination efforts. Moreover, unlike the fuel debris in the Cascade Wall, the radioactive fragments beneath the ground were not isolated from the environment and had been gradually contaminating the groundwater.

Actually, Goloskokov rarely used the word "fuel" during our conversation. Instead, he used the term "fuel-containing material," or FCM. Nuclear reactor fuel is in the form of pellets of slightly enriched uranium that are contained in zirconium tubes. The stuff inside the Sarcophagus was no longer in that form. From 10 to 36 tons of core fragments were thrown into the central hall and upper levels of the reactor building in the explosion. Some fuel and fission products melted in the graphite fire and mixed with the building's structural elements together with the extinguishing materials dumped on the reactor. In the bubbler pool—a safety system beneath the reactor hall—the FCM took the form of brown ceramics and in one of the rooms on the third floor, black ceramics. One corridor contained FCM melted together with metal while the elephant's foot in room 217 was known as an LFCM—a lavalike fuel-containing mass (Plate 8).

Most of the FCMs were far too radioactive for scientists to study in any detail. So all of the numbers and estimates of how much fuel they contain were very approximate. For example, the only hint of the elephant's foot's properties was obtained when someone standing at a relatively safe distance shot at the thing with a machine gun and broke off a piece that was retrieved remotely for laboratory study.

Goloskokov laughed again. "But this was just a tiny chip off a huge block. There's no way to know how representative a sample it was."

Only 25 percent of the premises were accessible to people. The rest were either physically blocked by debris or too radioactive to approach—with readings of up to 3,300 roentgens an hour! Just 2 minutes in such areas can bring on acute radiation illness, while 10 minutes can yield a fatal dose. Robots and remote sensors must be used to measure and monitor the premises.

One robot, named Pioneer, was built by the same Pittsburgh-based firm that built the spunky Sojourner that explored Mars in 1997. A tractor-driven, remotely operated contraption similar to a small bulldozer, Pioneer was delivered in 1999 but was never put to work because the Shelter's innards don't resemble Martian landscapes and the robot proved unable to cross even the smallest obstacle.

The robots that were used were developed by Ukrainian engineers. In 2004, four of them were trundling around the reactor's ruins, taking pictures and drilling samples. My favorite resembled a spider with radiation sensors and magnetic feet that let it climb vertically up metal walls.

Based on information from the robots, sensors, and samples, scientists think (but can't know for sure) that none of the FCMs have formed a critical mass capable of a self-sustaining fission reaction. On occasion, some sensors have picked up evidence of neutron activity—which would indicate fission—but other sensors didn't confirm the readings.

"So, we think that those were malfunctions," said Goloskokov.

There are also about 10 tons of radioactive dust covering all parts of the reactor building. Worse still, the amount of dust is growing.

"The Shelter is not hermetic and was never intended to be," Goloskokov explained. "There are about 100 square meters of cracks and openings."

These cracks serve as a way for dust to get out into the environment. Birds fly in, get sprinkled with dust, and then fly out, carrying the contamination great distances. Such vectors are minor, however. But if one of the large, unstable objects inside—such as the 200-ton core cover that was blown on its side and hangs over the empty reactor vault—falls, it would raise a significant cloud of radioactive dust that could drift out into the environment.

The cracks also allow precipitation to get inside. In the early postdisaster period, the FCMs were hot and glass-like, so the water merely evaporated on contact. But now that the FCMs have cooled, the water no longer evaporates. Shelter experts—and an entire interdisci-

plinary institute of the Ukrainian Academy of Sciences is devoted to
Shelter science—estimate that there are now from 2,000 to 3,000 cubic
meters of water inside the structure. In comparison, an Olympic-sized
pool holds 4,000 cubic meters.

This "unit water," as it is known, affects electrical and diagnostic
systems, corrodes metals, and deteriorates concrete. In the winter it
freezes, cracking FCMs and creating dust. Most dangerously, it also
leaches out water-soluble forms of enriched uranium and transuranic
elements such as plutonium and then trickles and flows into the
reactor's basements, where it is ankle-deep in places. And the amount
of transuranic elements in it is increasing with time. This means that
the water poses a nuclear risk.

In contrast to fission products such as cesium-137 and strontium-
90 that are merely radioactive, transuranic elements such as uranium
and plutonium are also capable of fission. Since water is a moderator,
slowing neutrons so that they are more likely to hit other atomic nuclei in a chain reaction, the transuranic soup accumulating in the bow-
els of the reactor poses the risk of starting an uncontrolled nuclear
reaction.

This is why pumping that radioactive water out for processing is a
top priority. It is also a huge undertaking. Although the plant has facilities for processing liquid waste created in the course of normal
reactor operations, it is not suited for the dangerous job of removing
the transuranic elements in the unit water. The problem is that when
radioactive materials are separated from water, they are usually concentrated to take up less space for storage. But concentrating transuranic elements could create a critical mass that will sustain fission. So
the decontamination facility must somehow ensure that such accumulations do not occur.

I swept my arm towards the Shelter, "So, will it ever be clean?"

"A green field?" Goloskokov smiled and shook his head. "I doubt
it. You see, the Shelter Object itself is radioactive waste that should be
processed and safely stored."

Goloskokov laughed at the notion of such a mind-boggling task.
"We're talking about hundreds of thousands of tons. That is more radioactive waste than exists in the entire world! Disposing of it is simply unrealistic."

CONFINEMENT

For security reasons, the only place that visitors can legally take close-up pictures of the Shelter is in a parking lot located about 250 meters away on its northwest corner, where there is a good view of the Cascade and Buttress Walls.

The structure is a gray and grimy eyesore surrounded by concrete partitions and concertina wire. Photographs barely convey how decrepit it looks in real life.

Rimma accompanied me to the Shelter. But I remembered Goloskokov's words in describing it. "The Shelter is a very. . . . " His voice had trailed off and he shrugged as if to underscore the inadequacy of his description: "It's a very *specific* object, built under extremely high radiation levels, often remotely. It's a covering to protect personnel and the environment from the high levels of radioactivity inside." In that light, it seemed unfair to criticize the structure, which has done a pretty good job given its extraordinarily difficult task and the unique circumstances of its construction.

My dosimeter beeped rapidly in the parking lot, reading around 650 microroentgens an hour depending on where I walked. Closer to the Shelter itself, hourly doses are around 200 to 300 millirem. They were as high as 15 *rem* on the roof! The only comparable radiation doses on the plant's ground were in the radioactive waste storage facilities where the average dose levels run to about 25 millirem an hour.

The parking lot was for the Shelter's Visitors Center, which has an excellent view of the Sarcophagus from its second-story windows. The view is so excellent, in fact, that for security reasons you are not allowed to take pictures of it. My dosimeter beeped alarmingly near the glass and the numbers kept climbing. They kept climbing for so long that I decided not to stand there waiting for the final tally and left the device on the windowsill for a few minutes. A digital sign on the wall informed us that radiation levels on the building's roof—where they reach their maximum at the Visitors Center—were 1.58 milliroentgens an hour and when I picked up the dosimeter a few minutes later, it displayed 600 microroentgens. But the reading dropped rapidly to 70 microroentgens just 25 feet away, near a TV set and VCR into which the center's guide Julia Marusich popped a cassette about how the Shelter was built.

It was an engrossing short documentary that finally helped me reconcile images of the ruined reactor building (Plate 1) and the Sarcophagus (Plate 7). The direction of the explosion was north, the narrator explained, punching a crater into the Cascade Wall. So that part of the Sarcophagus was built first, under such hasty and dangerous conditions—with radiation levels up to 2,000 roentgens an hour—that it is one of the least stable parts of the building today and requires reinforcement. But most parts are unstable because the explosion caused structural damage to the reactor building inside, although lethal radiation levels made it impossible to assess in the disaster's immediate aftermath.

Much of the Shelter was built in prefabricated pieces, placed on or above the reactor's remains, and many of them were just laid in place and held by friction alone. A lot of the pieces never fit together properly, and as the underlying rubble has shifted with time, so have pieces of the Shelter.

To those who complain about the structure's inadequacy, Goloskokov responded: "We did more than was possible and less than we wanted."

On a detailed scale model of the Shelter, Marusich swung open the buttress wall like a door to reveal the structure's astonishingly messy innards of what looked like metallic stalactites and tangled spaghetti. Tiny human figurines helped gauge the size of things and pink flags scattered about the premises indicated where sensors have been placed. She pointed out the reactor's tilting western wall.

"It is propped up by the Shelter's buttress wall, but it is unstable and could fall in an earthquake, knocking down the buttress wall and spilling radioactive debris and fuel-containing materials," she said.

Although reverberations of Crimean and Carpathian earthquakes can be felt there, Polissia has not traditionally been considered a seismically active region. Nevertheless, the Chernobyl station stands at the intersection of several faults. In fact, there is some evidence that a small earthquake shook an area less than 10 miles from the nuclear station at 1:23 a.m. on the night of April 26, 1986, and may have contributed to the disaster. Unfortunately, no one can say for sure because there were no seismic stations anywhere near Polissia. And there aren't to this day.

The western wall will be stabilized with two towers outside the buttress walls. A supporting column on the southern wall must also be reinforced, along with southern parts of the roof.

While critically important, such stabilization works will merely help keep the Shelter from collapsing. Large posters in the Visitors Center explain what needs to be done to make it environmentally safe, while flags on the walls symbolize the consortium of 26 donor countries that are supposed to help foot the estimated $768 million bill for what is officially known as the Shelter Implementation Plan, or SIP. In the bureaucratic jargon of its public relations materials, the SIP is supposed to "provide a decision based route to choosing technical options without, at this stage, defining the ultimate technical solution."

In plain English, this means building a new covering over the old leaky Shelter Object and then, maybe, hopefully, eventually—but not necessarily and definitely not within the SIP budget—figuring out what to do with what's inside. Not surprisingly, the approach has led to some angry divisions between critics, who complain that the SIP merely means to sweep Chernobyl under an expensive rug without actually disposing of its contaminated insides, and proponents, who respond that the problem simply can't be solved with current technology and budgets, so the best thing to do is to better isolate the radioactive mess from the environment. Critics also charge that the structure won't be hermetically sealed against moisture and does nothing to prevent the moisture that is already inside the Sarcophagus from condensing. Proponents respond that it is better than having 100 square meters of cracks and openings like the current covering.

After considering a number of different ideas for a new shelter, the SIP experts settled on a giant arch that will be 328 feet high, span 853 feet, and weigh 20,000 metric tons. Officially known as the New Safe Confinement, it is supposed to last at least 100 years, although it may last as long as 300 if it is properly maintained.

To be assembled at a relatively safe distance from the Shelter and slid into place when it is complete, the Confinement is being billed as perhaps the largest movable structure ever to be built (Figure 4).

After removing the contaminated topsoil around the Shelter to reduce workers' radiation exposure, a concrete and Teflon foundation about 10 feet thick and 50 feet wide will be built parallel to the Shelter's northern and southern walls. The arch's components will be preassembled in a shop in sections that are as large as possible—some may be as much as 200 feet long and weigh up to 200 tons—to reduce

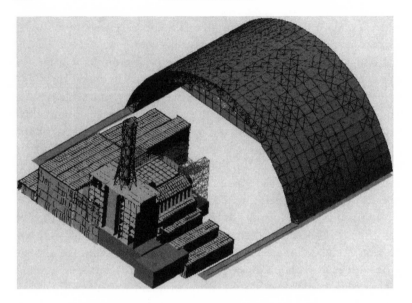

FIGURE 4 The New Safe Confinement.

the amount of time spent at the assembly site that will be set up in the shadow of the buttress wall.

As each section of the arch is completed in the assembly area, it will be slid along the foundation's Teflon surface towards the Shelter, freeing the assembly area to build the following section. As new sections are slid towards their predecessors, those will in turn slide closer to the Shelter. After all the sections are built, hydraulic pulleys will slide the arch into place over the Shelter—a process that is expected to occur in 2008 and take less than a day.

This, at least, is the plan. But the Confinement is a complex object that poses a radioactive danger, and much must be done to prevent overexposure of the workers who will build it. In fact, as of 2004, there was still some debate as to whether an arch was the best form for a new Chernobyl cover. Some Ukrainian scientists expressed concern that it is simply too heavy for Polissian terrain, which is pocked with saucer-like depressions that suggest complex geological processes going on under the surface. Confinement planners haven't studied them at all.

While waiting for the new Confinement, Shelter experts have tried to protect the FCM from water and simultaneously fix the dust with an experimental polymer called EKOR. Billed as radiation- and corrosion-resistant material designed to maintain its integrity for centuries, EKOR is a white foam that hardens and seals the FCM. But because it hasn't actually been tried outside the laboratory for any length of time, it is being tested on a pile of FCM slag in one of the bubbler pools to see how radiation resistant it really is.

Once completed, the arch will be like a nuclear Matrioshka doll, with the Confinement on the outside, the Shelter inside the Confinement, the ruined reactor inside the Shelter, and the radioactive FCMs inside all of it. Designed to withstand a Richter 6 earthquake and a Class 3 tornado—but not a plane crash because the Chernobyl area is a no-fly zone—the arch is supposed to keep the radioactive materials in and keep most moisture out, eventually eliminating the need for EKOR, although the experiment on the slag pile may still prove useful for the planet's other radioactive messes.

The Confinement's interior will be rigged with remotely operated cranes and other equipment for dismantling the upper parts of the Shelter and ruined reactor building. At that point, future nuclear archaeologists might find it possible to dig up the pile of extinguishing materials in the central hall and figure out how much missing fuel lies beneath.

Ideally, all of the highly radioactive insides should then be scooped, scraped, and swept up into containers and safely stored. But this is a gargantuan task, given the risks of occupational radiation exposure and the diffused location of the stuff—including all of the dust— throughout the reactor building. Moreover, since the melted fuel melded together with various reactor construction materials, it is no longer a matter of cleaning up what remains of the 200 tons of fuel, but of disposing of 3,000 tons of fuel-containing materials totaling about 20 million curies of radioactivity. The Confinement project budgets some money to research possible ways to safely dispose of the FCMs. But there is no money—and, currently, no realistic way—to actually do it. No one even expects to start for another 40 or 50 years.

By then, perhaps, scientists will figure out how and where to store the stuff. The FCMs are highly radioactive, long-lived waste that can't be stored in the same way as the low- and medium-active, short- and medium-term waste stored on the plant's grounds and in the barrows of Burakivka that I visited at the start of this journey. One storage option is to bury the stuff deep in the earth somewhere within the borders of the zone, but scientists have only begun preliminary reconnaissance work to identify geologically appropriate sites. Besides, geological storage is too expensive for Ukraine right now.

Even if (or when) the FCMs are cleaned out, the radioactive Sarcophagus itself will probably remain inside the Confinement for as long as both can stand.

What happens after the Confinement's maximum three-century integrity expires is anyone's guess. The scientists and engineers at the Chernobyl station didn't seem eager to speculate about questions with no practical answers.

I decided to go talk to Volodymyr Kholosha, the head of zone administration, about it. A youthful 49-year-old who trained as a power engineer and worked at the nuclear station at the time of the disaster, Holosha occupied a sparely furnished office decorated with radiation maps on the second floor of Chornobyl's former Soviet town hall, a few blocks away from Chernobylinterinform.

When I arrived, he invited me to sit and looked pointedly at his watch to indicate that I had little time, so with few preliminaries I asked him about the zone's future.

With the assurance of someone who has done a good deal of thinking on the subject, Kholosha explained that some areas—namely the grounds of the nuclear station and the adjacent areas—will be used for decommissioning the three undestroyed reactors, processing and storing radioactive waste, and maintaining the Shelter Object and eventually the Confinement. Territory is needed to do all that. For example, 60 hectares in the Red Forest have been decontaminated to build a nuclear waste facility called Vector.

About six square miles containing the region's original flora and fauna, or rare species, will be set aside. Already mapped and surveyed, they will be like radioecology sanctuaries, where nature will be allowed to renew itself and rehabilitate without human interference.

Then there were what Kholosha termed "priority rehabilitation" lands—territories, primarily in the southeastern parts of the zone where people can work with minimal radiological restrictions. He told me that saplings for future forestry farms had already been planted there and the zone was also considering fish farming, to breed young fish that can be grown to adulthood in other places.

"And what will the zone look like in 100 years?" I asked.

Kholosha sighed. "The zone will be a green oasis."

He stopped as if to check off parts of its current landscape in his mind. "There will be no Chernobyl nuclear power plant. It will be decommissioned and deconstructed. And there will either be a new Confinement, to replace the one we are about to build. Or there will be no Confinement at all because its contents will be cleaned up."

When I raised a skeptical eyebrow at that optimistic prediction, Kholosha smiled: "We can't know what new technologies will be developed by then."

"There will be some nuclear waste storage facilities around the territory," he continued. "There may be more people living here than there are today, especially on the periphery—but not many. There will be fewer people working in the zone administration and at the nuclear waste facilities than there are today. And there will be many, many more animals."

"I know that this is more in the sphere of science fiction. But what do you imagine 300 years from now?" I asked.

Kholosha shrugged. "That's hard to predict. Maybe the number of people will grow, though I don't think there will ever be as many people as there were before the accident. And they'll probably be peculiar characters, like those people you have in America who go off to live alone into the wilderness." I couldn't help but think of Ted Kaczynski, the Unabomber, who holed up in his Montana cabin to write antitechnology manifestos during breaks from his bomb attacks. It was a chilling thought for the zone's future.

The idea of repopulating the zone as the radionuclides decay with time is a subject of hot dispute. As of 2004, contamination levels were safe enough to lift restrictions on about 150 square miles. The clean areas will grow with time as more and more cesium and strontium decay. With the passage of 30 years, which is equivalent to the half-life of cesium-137 (30.7 years) and strontium-90 (28.79 years), restrictions can be lifted on more than half of those parts of the zone contaminated with only those radionuclides. Not all of it will be clean enough for permanent residence, but even the parts that can't be inhabited can be used for some kind of economic activity.

Proponents of lifting restrictions on parts of the zone have come up with various ideas for what kind of economic activity can actually be conducted there even today. In Belarus they've taken up horse breeding on the cleaner periphery of the radiological reserve, an idea that is also being considered in Ukraine. Other suggestions include fur farms and beekeeping, not for the radioactive honey produced but for breeding new colonies for sale.

A "Made in Chernobyl" label would probably be a tough sell, although aggressive marketing might achieve wonders. More seriously, the zone's contamination is a constantly shifting kaleidoscope. High winds, hurricanes, and forest fires can spread radionuclides to clean zones. While such conditions can spread contamination outside the zone's borders as well, the risks are greater close to the most highly contaminated areas.

When I asked about repopulating the zone, Kholosha responded: "That is less a question of radiology than of sociology, psychology, and economics. We're not talking about professionals who work here, but of civilians returning for permanent residence.

"Imagine what's needed for this to happen. First of all, the infrastructure would have to be renewed. After 18 years, everything has crumbled—stores, schools, post offices, sewage systems. But let's say that money is found to do that. Imagine a child growing up here. He can't go fishing because the fish in ponds and lakes are too radioactive. No matter what direction he walks in, he'll eventually encounter barbed wire. And all this time he'll grow up in the shadow of the Sarcophagus. What will be the psychological impact on this child's life?

"It just doesn't make sense to resettle the zone when there are uncontaminated parts of Ukraine that could use more people," said Kholosha. "Sure, you could do it as an experiment. Let people with children come to live here, ship in clean produce. But I don't think it's a good idea."

The zone is a sink for radioactivity, and this poses a radiological danger in and of itself. But it also acts as a barrier between the radiation and human exposure because its checkpoints keep a distance between the two. Letting people return for permanent residence would remove that barrier.

So, given those arguments, who was actually in favor of repopulating the zone? In Belarus the government was encouraging resettlement in villages on the border of the reserve, but not directly inside it. In Ukraine it was a grassroots movement.

"The *samosels* want it because they want a more normal situation. And evacuees who grew up with the climate here and can't get used to living in, say, the black earth steppe region want to return," Kholosha told me. "These ideas are not coming from the top at all."

Kholosha continued: "But it really is not a good idea. We analyze the things the *samosels* eat. Milk from cows pastured outdoors exceeds strontium limits, as do vegetables and fish."

This poses a problem outside the zone as well. *Samosels'* family and friends can apply for permission to enter the zone, increasing the number of potential poachers. Just a few weeks earlier, a businessman I knew boasted about visiting someone in the zone, going fishing, and then taking the fish home to his family in Kiev.

"Someone checked it with a dosimeter and it was fine," he had explained when I asked if he wasn't concerned about radioactivity. He was quite perturbed when I told him that the dosimeter only measured gamma radiation in the air and couldn't tell him anything about the amount of radioactivity inside the fish. But at least he knew where the fish came from. Some poachers sell zone fish or game at markets, and no one is the wiser about their origins.

The town of Chornobyl's *samosels* pose a special problem. First of all, there are many of them. About one-third of the zone's permanent residents live there. And because the town is the headquarters of the administration and its 3,200 employees, its infrastructure is maintained quite well, creating an illusion of normality that fuels the grassroots demand for legal resettlement.

Nevertheless, if new settlement is not allowed, the *samosels'* demands will gradually die with them.

Kholosha glanced at his watch and I rushed to pose my last question.

"And Pripyat?" I asked. "What will happen to Pripyat?"

Kholosha smiled sadly. "Pripyat will be ruins, like one of those Aztec or Inca cities."

WORMWOOD FOREVER

The previous year's old brown leaves skipped across Pripyat's crumbling central plaza in the mild spring breeze. Pripyat was so young when it died that the plaza never did get an official name. But it was the only part of the town that hadn't succumbed completely to forest, largely because it had been cleaned up when U.S. Vice President Al

Gore came to visit in 1998. It was also where buses carrying scientists and visitors inevitably parked, so what did grow usually got squashed.

While I pulled my dosimeter out of my vest pocket, Rimma crouched down to inspect a young sprout of wormwood that had poked up through the asphalt cracks. After our first botanical tour of Pripyat in Chapter 1, Rimma had become an experienced wormwood spotter. She plucked off a tiny leaf and crushed it for me to smell the familiar varnishy aroma.

Waist-high radiation levels were 36 micros an hour, but they were higher closer to the ground because that's where nearly all the radionuclides are concentrated. Although it couldn't actually measure radionuclides inside the wormwood, which was accumulating cesium as its spring juices activated, the dosimeter registered 120 micros in the air around it.

The plaza was a patchwork of readings. Near a plug of dried moss, they were 160; on some crumbling asphalt, 200. On another spot of moss the dosimeter beeped rapidly and kept climbing to finally hit 700 microroentgens an hour.

"That's even higher than near the Sarcophagus," Rimma noted. "The graveyard here is especially radioactive. Radiation levels are in milliroentgens in some places there."

Placed between the town's barbed wire enclosure and the Chernobyl station, the graveyard had been directly in the path of the debris from the initial explosion. But since I had been in so many "especially radioactive places" in the course of my journey, I decided that my story could do without more of them, especially with a dosimeter that was worthless in any place where radiation levels were higher than two milliroentgens an hour.

In fact, I didn't even know the total dose I had received in researching this book. I could have easily purchased one of the clip-on dosimeters that all zone workers wear to measure their annual doses. But I started going to Chernobyl and collecting material for newspaper articles long before I knew I'd be writing a book. I rationalized this lapse on my part with the fact that Rimma goes to all the places I had visited and far more frequently. So, if she hadn't exceeded dose limits then I probably hadn't either.

I did, however, calculate my approximate exposure by adding up the number of hours I spent in the zone, multiplying that by the average zone radiation level of 43 microroentgens an hour plus adding bonus exposure based on my time spent in so-called especially radio-active places such as the left bank polder, the Red Forest, and the shadow of the Sarcophagus. The number I came up with was a total exposure of about 25 milliroentgens. According to nearly every national and international radiation protection standard, this was well below the maximum recommended annual exposure. But as my journey drew to a close, I decided that whatever my total dose actually was, it was fast reaching "enough."

Near some wormwood growing on a knoll of dried grasses, the hourly level was 600 microroentgens, but quickly dropped back to double digits when I straightened up to stroll amid the ghostly high-rises. Even after 18 years, when babies born in 1986 had reached adulthood, Pripyat was too radioactive to be inhabited. Its buildings had so crumbled that it seemed unlikely anyone but a desperate fugitive would try to live in one of them even if it wasn't contaminated.

Indeed, the town looked even more decrepit than it did when I visited it at the start of this story. The grammar school was flooded with water from several days of rain. Water was still trickling through the old coatroom where an abandoned bird's nest was perched on one of the clothes hooks. Chunks of peeling paint curled on the walls, and spongy linoleum squished beneath our feet as we carefully made our way down a corridor. The wood had rotted away in places, and a few times I stumbled into holes and sagging depressions in the floor. It did not feel structurally sound at all, and we quickly emerged into the tangled overgrowth outdoors where large clusters of red firebugs scurried about the ground in their spring orgies of feeding and mating.

The town's erstwhile rose gardens, lawns, and tidy tree-lined promenades were succumbing to the infertile soil's natural, hardy *Artemisia* vegetation. The Chernobyl region was once again living up to its medieval name and Kholosha's predicted green oasis was well on its way to becoming a reality. Indeed, although they don't often say so outright, it seemed to me that opponents of resettling the zone like the idea of its remaining a wildlife sanctuary.

James Lovelock, the inventor of the Gaia theory of Earth as a kind of living superorganism and a proponent of nuclear energy as a "green" alternative to climate-warming fossil fuels, recently cited Chernobyl's transformation into a wildlife park in a British newspaper article he wrote in support of nuclear energy. He even proposed the deliberate creation of Zones of Alienation: "I have wondered if the small amounts of nuclear waste from power production should be stored in tropical forests and other habitats in need of a reliable guardian against their destruction by greedy developers."

It's a wacky idea with a certain logic: The only way to save the planet from ourselves is to deliberately create involuntary parks, unnaturally natural places that are just fine for wildlife but too dangerous for humans.

The success of Lovelock's scheme would depend partly on what the developers were developing. If it is a condominium complex for rich folks, such a radioactive sword probably would keep potential buyers out of that particular patch of tropical Eden. But I somehow doubt if nuclear waste would stop the logging of tropical forests, especially if the lumber is clean and if poor Third World workers are doing the work and absorbing the radioactive background doses.

Still, the benefits of nature conservation alone are not a very convincing argument in poor countries. With the help of impoverished peasants who call the butterflies "worms," mobs and mafias have been illegally logging the ancient Mexican evergreen forest that is the winter haven for the monarch butterfly, and Mexican law has been helpless to prevent them.

Of course, aside from being risky, unethical, and way too weird for serious consideration, the problem with using radioactivity as a keep-out sign is its invisibility. And this very invisibility is what makes more graphic keep-out signs crucial in radioactively contaminated areas. But some radionuclides are so long-lived that the signs will either disintegrate or become unintelligible long before they become unnecessary.

This very problem fueled a bizarre American debate in the 1990s, after the federal government authorized the country's first long-term radioactive waste storage facility—the Waste Isolation Pilot Plant, or WIPP—in southwest New Mexico. To eventually store drums of contaminated, low-level waste such as clothes, tools, rags, and solid residues, the Department of Energy mined more than 10 miles of chambers 2,100 feet beneath the surface in the salt layers left when an ancient sea evaporated. Since geological activity at the site is deemed unlikely to disturb the waste once it has been stored (the timing of which, as of this writing, is unclear), the danger lies in its penetration by people unaware of the dangerous radioactive stuff inside or by people such as terrorists, who know full well what's inside and want to get some for a dirty bomb or other nefarious device.

For the foreseeable future, intruders are supposed to be warded off by the warning signs, barbed wire, and checkpoints that surround any highly secured place. But what to do about the unforeseeable future? Since some fission products are about as permanent as the landscape itself, environmental regulations require that public warnings about WIPP must remain effective for 10,000 years! And because even the U.S. government institutionally recognizes that it is not eternal, the warnings must work passively, without further human intervention, after the first century.

Of course, 10,000 years is an arbitrary time period. Much of the waste will still be radioactive after that. Nevertheless, it is far beyond the boundary of humanity's poor collective predictive powers. For a glimpse of what this means, 10,000 years ago the seeds of civilization were just being planted with the invention of agriculture in the Middle East. The first cities arose in Mesopotamia about 5,000 years later. Given the pace of technological development today—as well as the destructive uses to which this technology has too often and too sadly been put—it is impossible imagine what 10,000 years into the future will hold. More accurately, it is possible to imagine but impossible to predict.

So, how to design warning signs that not only will last 10 millennia but also will be intelligible 400 generations from now? After 20 to 40 generations, any language becomes gibberish to its speakers' descendants. Read Beowulf to see how the English language has changed since the eighth century. Even sixteenth-century Shakespeare can be tough going in parts. And for anyone who believes that the current Information Age will preserve dictionaries and grammars forever, try retrieving a document stored on an original "floppy" disk without going to an antique computer collector to find a drive that can read it.

But even if the carrier of the message remains intact and legible, what kind of marker will preserve the meaning of "keep out" to people for whom these two words will be like the still-undeciphered Linear A script of ancient Crete? Even the meaning of the traditional trefoil sign for radiation, designed in 1946, could change with time—like the swastika, which symbolized the sun and good fortune in many cultures before the Nazis contaminated it.

Still, if the far-into-the-future New Mexicans understand the message, there is no reason to think they would believe it. Vast halls of the finest museums would echo with emptiness if the world's Indiana Joneses believed in mummies' curses. To the contrary, a mummy's curse is like an X marking the spot where treasure might be found.

Arizona State University geographer Martin Pasqualetti wrote about the problem of how to "alert the prudent without attracting the foolish." He suggested that a "landscape of illusion"—one that would leave the WIPP site anonymous and unmarked on the surface but with subterranean warnings to anyone getting too close to the radioactive stuff inside—would provide better protection than what he called "landscapes of repulsion" that would be just as likely to attract the curious.

Design contests for so-called universal warning signs illuminate the dilemma. The signs, whose message is not based on words, have been proposed for WIPP as well as for another waste repository planned for Yucca Mountain in Nevada. The winning entry in one competition proposed planting the site with genetically engineered cobalt blue cactuses, despite the fact that such an unusual and alluring landscape would probably attract people rather than ward them off. The same could be true of more foreboding earthworks such as the "Landscape of Thorns," composed of 50-foot-high concrete spires with sharp points jutting out at all angles, or "Forbidding Blocks"—giant, irregular blocks of black stone too closely spaced and hot to provide shelter.

If you were in the neighborhood in the year 10,000 and heard about such a mysterious, ancient monument, wouldn't you go take a look? I would. Even anti-art or anti-architecture can't shout a danger warning without attracting the very people it intends to repel.

The U.S. Department of Energy has—perhaps wisely—settled on a more mundane design for WIPP, consisting of a 33-foot-high earthen hill covering 120 acres and bordered by giant granite warning monuments. A roofless granite information center engraved with written and pictorial messages will sit in the center of the hill, and archives about the site's contents will be stored in various locations around the world. But building will not begin until the end of the twenty-first century, so future generations still have time to come up with better ideas for warning their own descendants about the radioactive legacy our generation bequeathed them.

Thus far, however, no one is planning to inscribe semiotic warnings on the Confinement arch that will cover the Chernobyl reactor, perhaps because its lifetime will be short enough for current languages to remain intelligible and for current security to remain intact. Yet even if the arch were to be gouged with trefoil hieroglyphics and warnings, nothing similar is even remotely possible for the radioactive wilderness around it and the contaminated kurgan burial mounds containing the remains of buried villages such as Kopachi, Yaniv, and Chistohalivka.

Aside from barbed wire, checkpoints, and presumably eternal governments, no one I ever spoke to in Ukraine or Belarus conceived of, or even thought about, a permanent way to keep people out of the zone for the 300 years that amount to about 10 cesium and strontium half-lives. This is to say nothing about the 10,000 years mandated for WIPP, much less the 24,110-year half-life of plutonium-239.

Was this irresponsible? Or was it hubris to believe it was even possible?

The first floor lobby of the Pripyat high-rise was a dark, treacherous passage smelling of mildew and littered with a deep pile of ripped mattresses, chunks of paint and plaster, old shoes, and dismembered dolls. The elevator lay dead in its shaft.

I followed Rimma up the stairwell, listening to the rubble from collapsing walls and ceilings crunch beneath our feet.

Built by largely inexperienced members of the Komsomol—or the Young Communist League—during summer vacations, the buildings were structural horrors. There were no real right angles and the concrete slabs that were the building's carcass didn't meet. The stairwell was built in a chamber up the side of the building, but the exterior wall inexplicably ended six inches from the floor, letting in the outside air. The stupidities committed under the Soviet system—and its ability to creak along for so long despite them—are still astonishing. The poor construction will never last as long as the Egyptian pyramids or Aztec cities.

Looters had taken anything of value. Chernobyl coffee table books with photos taken soon after the disaster show intact furnishings and neatly made beds. But Rimma showed me a piano in a flat on the fourteenth floor. It was missing most of its wood, so its innards were exposed, and the scale I played on it was horribly out of tune. It and an overturned table scattered with some books and paper were the only things left in the one-room apartment.

"This is the only place in the world where you can see what the Soviet Union was like before perestroika," said Rimma.

And it was a very strange place, indeed.

One of the books on the floor was the *Short Course on the History of the CPSU*. The CPSU was the Communist Party of the Soviet Union, and the *Short Course* was its sacred text. Evidently, no looters found the book worth taking.

The floor was littered with many pieces of paper curled up like loose cigarettes, and I unrolled a few to find faded black-and-white photographs of babies, families, and a man squinting into the sun and holding up a large fish against a backdrop of tall reeds.

A yellowed copy of *Izvestia*, dated January 29, 1985, proclaimed in large block letters: "We won't let the world detonate!" According to the article, "Soviet youth" said this in "hot support" of some utterance about world peace from the geriatric and generally useless General Secretary Konstantin Chernenko, who replaced Andropov and was the last of the old-guard Soviet Politburo assembled by Leonid Brezhnev. It was one month into his tenure when Chernobyl's No. 4 reactor started generating electricity. Mikhail Gorbachev took his place in April of 1985.

The kitchen was empty but for the cabinets. A 1986 calendar displaying Kiev's World War II museum was pasted on one of the cabinet doors. All of the red-letter days had long since faded away in the sun, leaving ghostly nondays—the nondays of Pripyat after its people left.

We left the apartment and climbed up the last two floors to the blacktop roof, where there was a panoramic view of Pripyat's empty high-rises and the Shelter Object looming in plain view. It was probably the same high-rise that the anonymous KGB cameraman climbed to film the smoldering reactor in the silent movie I watched in Kiev's Chernobyl museum at the start of my journey.

I recalled the shock waves when news about a Soviet nuclear disaster first leaked out of radiation detectors in Sweden. For months afterwards, Western media were filled with wild speculation and worst-case scenarios that tried to fill the vacuum of Soviet silence, while satellites tracked the radioactive cloud that drifted about the globe. With its invisible menace of radiation, the disaster touched the entire northern hemisphere and all the fears of the nuclear age.

I tried to picture how the shiny new arch (or whatever form the new Confinement eventually takes) would look on the horizon. Like the radioactive waste barrows of Burakivka, the Confinement arch should someday be painted green on the zone maps because the stuff inside it will, at least for the near future, be relatively safely enclosed. But the soil, plants, and animals of the entire Zone of Alienation are also, in a way, fuel-containing materials. And no confinement arch is big enough to isolate them from the environment. In the zone, FCMs *are* the environment.

Like the future WIPP monument built intentionally to communicate a cultural message to posterity, the Confinement arch would be like Stonehenge, the Egyptian pyramids, or their Eurasian counterparts, the earthen kurgans that dot the steppes—transcending the values of a particular culture and speaking to all humanity. While the arch itself is not intended to last as long as Stonehenge, in the absence of scientific breakthroughs in radioactive cleanup, some kind of protective shelter will have to stand on that spot of Earth for far longer than the megaliths, pyramids, and kurgans combined.

Will these successive shelters through the ages speak to all of humanity? And more importantly, what will they say? The tomb over the ruined fourth reactor was like a monumental Rorschach test, perhaps more revealing about the person looking for meaning in it than about the thing itself. Was Chernobyl's message one of hazard, about the dangers of technology and the fact that all of us, 5 billion strong, live downwind from 300 nuclear reactors that are operated by mere people and have a statistical probability of one meltdown every 30 years?

Or was its message one of hope, that no matter how humanity messes up, nature will persevere—even if it is forever changed and unnaturally natural, like the radioactive landscapes of Chernobyl? Perhaps the arch would one day become a kind of environmental shrine, eerily sanctifying the radioactive wilderness around it. Or would it desecrate that wilderness? I imagined future philosophers making pilgrimages to contemplate the shelter shrine and come up with answers.

I didn't have any.

Instead, I remember a dream I had soon after the disaster. In it I was out on the town with a group of faceless strangers on a dark, gray night. Suddenly a staircase appeared in a kind of scaffolding and I climbed it into my parent's brightly lit kitchen. In the middle of the linoleum sat the Chernobyl reactor core. It wasn't big and looked like nothing special, but I knew it was emanating deadly, invisible rays. We walked over and around it in the kitchen, going about our business and pretending it wasn't there. But I knew it was dangerous and told my mother: "We have to get rid of this thing!" And her eminently sensible response was: "But where are we supposed to put it?"

In my hyphenated Ukrainian-American existence, "America" was the streets and schools, bookshelves and television. But "Ukraine" was family, and the family gathered mostly in the kitchen. In dream logic, Chernobyl was in Ukraine, Ukraine was in my parents' kitchen, and therefore Chernobyl was in my parents' kitchen.

The worst-case scenarios depicted a dead land, a black hole that had moved me in an immediate way, as though I had lost a part of myself. In the years that followed, the disaster became a kind of hobby for me. I studied all Chernobyl matters closely and was convinced that the Soviets were lying about it even after Gorbachev ushered in glasnost. In fact, they did lie about a great deal. But by the time such details emerged in the early 1990s, I was living in Kiev where, oddly enough, my Chernobyl fascination suddenly dissolved.

I never understood why until I stood atop that Pripyat high-rise, surrounded by the wormwood forests, swamps, and fields blooming in the pastel palette of early spring. If a nuclear disaster really *is* in your kitchen—or in your metaphoric backyard—and there is nowhere else you can put it, it seems best not to think about it too much. Not, at least, until many years have passed, and the bountiful evidence of nature's nearly miraculous resilience and recovery makes the thinking more bearable.

Acknowledgments

In writing this book, I have e-mailed, phoned, interviewed, and interrupted the lives of dozens of people in a variety of disciplines and on several continents. In fact, I never learned the names of many of the zone employees—the checkpoint guards, drivers, canteen cooks—whose services and kindness helped make every single trip there memorable. But I would like to thank the following people for contributing to this book by agreeing to be interviewed, escorting me around the zone, responding to queries, providing scientific papers and background material, and/or reading and commenting on parts of the manuscript: Valery Antropov, Oleksandr Berovsky, Svitlana Bidna, Yuri Bondar, M.D. Bondarkov, Kate Brown, Ivan M. Bulavik, Igor Chyzhevski, Tom Hinton, Vitaly Gaichenko, Sergei Gashchak, Marvin Goldman, Oleg Goloskokov, Dmytro Grodzinsky, Volodymyr Kholosha, Sergei Koshelev, Lina Kostenko, Oleksandr Kovtunenko, Askold Krushelnyckyj, Ronya Lozynsky, Lydia Matiaszek, Sergei Mosyakin, Petr Palytayev, Peter Raven, Victor Riasenko, Mykola Rupashchenko, Valery Rybalka, Sergei Saversky, Wayne Scott, Olia Senyuk, Vyacheslav Shestopalov, Maria Shevchenko, Marian Sikora, Bruce Sterling, Andriy Sverstiuk, Genn Saji, John Thomas, Mykola Tkachenko, Ward Whicker, John Wills, Richard Wilson, Denis Vishnevsky, Natalia Yasynetska, Oksana Zabuzhko, and Tatiana Zharkikh.

Thank you to the staff at Chernobylinterinform—especially Rimma Kyselytsia and Maryna Poliakova—for their unfailing accommodation of my sometimes strange requests and to the *Los Angeles Times*, which published many of the articles that eventually led to the idea for this book.

I would also like to acknowledge the following sources for the scientific material in the narrative. Any mistakes or misunderstanding of their content are entirely my own fault. V. H. Baryakhtar, ed. *Chernobyl: Zone of Alienation*, Naukova Dumka (Kiev, 2001) (in Ukrainian). *The Bulletin of the Ecological State of the Exclusion Zone*, Ukrainian Ministry of Emergencies and the Administration of the Zone of Exclusion and the Zone of Unquestionable (Mandatory) Resettlement (in Ukrainian). *The Human Consequences of the Chernobyl Nuclear Accident: A Strategy for Recovery*. A report commissioned by the UNDP and UNICEF with the support of UN-OCHA and WHO (New York, 2002). Karaoglou et al. *The Radiological Consequences of the Chernobyl Accident: Proceedings of the First International Conference, Minsk, Belarus, 18 to 22 March 1996*. The European Commission, Directorate-General XII Science, Research and Development (Luxembourg, 1996). Zhores Medvedev. *The Legacy of Chernobyl* (W.W. Norton & Co., New York, London, 1990). National Academy of Sciences of Ukraine. *Atlas of Chernobyl Exclusion Zone*, Kartohrafia (Kiev 1996).

I am grateful to everyone at the Joseph Henry Press who helped bring this book to life, especially Jeffrey Robbins, who acquired the manuscript.

Special thanks to my agent, Andrea Pedolsky, who believed in me.

Index